MW00412144

Macromolecular Symposia publishes lectures given at international symposia and is issued irregularly, with normally 14 volumes published per year. For each symposium volume, an Editor is appointed. The articles are peer-reviewed. The journal is produced by photo-offset lithography directly from the authors' typescripts.

Further information for authors can be found at http://www.ms-journal.de

Suggestions or proposals for conferences or symposia to be covered in this series should also be sent to the Editorial office (E-mail: macro-symp@wiley-vch.de).

Macromolecular Symposia:
Annual subscription rates 2005
Macromolecular Full Package: including Macromolecular Chemistry & Physics (24 issues), Macromolecular Rapid Communications (24), Macromolecular Bioscience (12), Macromolecular Theory & Simulations (9), Macromolecular Materials and Engineering (12), Macromolecular Symposia (14):

Europe	Euro	7.088 / 7.797
Switzerland	Sfr	12.448 / 13.693
All other areas	US$	8.898 / 9.788

print only **or** electronic only / print **and** electronic

Postage and handling charges included. All Wiley-VCH prices are exclusive of VAT. Prices are subject to change.

Single issues and back copies are available. Please ask for details at: service@wiley-vch.de

Orders may be placed through your bookseller or directly at the publishers: WILEY-VCH Verlag GmbH & Co. KGaA, P. O. Box 10 11 61, 69451 Weinheim, Germany, Tel. +49 (0) 62 01/6 06-400, Fax +49 (0) 62 01/60 61 84, E-mail: service@wiley-vch.de

For **copyright permission** please contact Claudia Jerke at:
Fax: +49 (0) 62 01/6 06-332, E-mail: cjerke@wiley-vch.de

For USA and Canada: Macromolecular Symposia (ISSN 1022-1360) is published with 14 volumes per year by WILEY-VCH Verlag GmbH & Co. KGaA, Boschstr. 12, 69451 Weinheim, Germany. Air freight and mailing in the USA by Publications Expediting Inc., 200 Meacham Ave., Elmont, NY 11003, USA. Application to mail at Periodicals Postage rate is pending at Jamaica, NY 11431, USA. POSTMASTER please send address changes to: Macromolecular Symposia, c/o Wiley-VCH, III River Street, Hoboken, NJ 07030, USA.

Macromolecular Symposia 222

Polymer–Solvent Complexes and Intercalates V

Lorient, France
July 11–13, 2004

Symposium Editor:
Y. Grohens, Lorient, France

pp. 1–296 · March 2005
ISBN 3-527-31325-7

Macromolecular Symposia

Articles published on the web will appear several weeks before the print edition. They are available through:

www.ms-journal.de

www.interscience.wiley.com

Polymer–Solvent Complexes and Intercalates V
Lorient (France), 2004

Preface
Y. Grohens

Author Index

Preface

This issue of *Macromolecular Symposia* contains the major contributions of the scientists who attended the 5th. International Conference on Polymer Solvent Complexes and Intercalates which was held in Lorient on July 11–13, 2004. This conference, organised by the Laboratoire des Polymères, Propriétés aux Interfaces et Composites (L2PIC) of the Uniiversité de Bretagne Sud in Lorient (France) under this auspices of the Groupe Français des Polymères (GFP), was a part of a series following the previous events in Meyrueis (1996), Ischia (1998), Besançon (2000) and Prague (2002), the proceedings of which were published in previous issues of *Macromolecular Symposia*. The aim of the conference was to bring together young and senior workers in the field of polymer solvents interactions, complexes and intercalates who use different techniques and approaches. One hundred participants from 20 countries took part in this meeting divided in four sessions, namely, biopolymers and gels, intercalates, specific phases and phase transitions and applications of polymer solvent complexes. There were 11 invited lectures, 25 oral communications and 32 posters.

Among the presentations, biopolymers and synthetic polymer gel structures investigated with various techniques were discussed. Intercalates, liquid crystals and polymers at interfaces were also included during these three days. Lastly, progress in DSC, neutron scattering, rheology, infrared and Raman spectrocopy results were presented. Very exciting discussions took place after each presentation and during the coffee breaks. The poster sessions were also good times for stimulating exchanges between researchers of the different countries. A guided visit of the India Company Museum in Port Louis was also a informal time for extra work discussions.

Summary of the lectures and posters were published in the program booklet and full texts of the major lectures and posters are published herein. This issue of *Macromolecular Symposia* provides a high quality overview of the up-to-date problems, new findings and forecast applications in this scientific field.

Y. Grohens

Phase Separation in Aqueous Polymer Solutions as Studied by NMR Methods

Jiří Spěváček

Institute of Macromolecular Chemistry, Academy of Sciences of the Czech Republic, 162 06 Prague 6, Czech Republic
Fax: +420 296809410; E-mail: spevacek@imc.cas.cz

Summary: Some possibilities of ^{1}H NMR spectroscopy in investigations of structural-dynamic changes and polymer-solvent interactions during the temperature-induced phase transitions in aqueous polymer solutions are described. Results obtained recently on D_2O solutions of poly(vinyl methyl ether) (PVME), poly(N-isopropylmethacrylamide) (PIPMAm), negatively charged copolymers of N-isopropylmethacrylamide and sodium methacrylate, and PIPMAm/PVME mixtures are discussed. A markedly different rate of dehydration process in dilute solutions on the one hand, and in semidilute and concentrated solutions on the other hand, was revealed from ^{1}H spin-spin relaxation measurements.

Keywords: aqueous polymer solutions; ionized copolymers; NMR; poly(N-isopropylmethacrylamide); poly(vinyl methyl ether); temperature-induced phase separation

Introduction

It is well known that some acrylamide-based polymers, including poly(N-isopropylacrylamide) (PIPAAm), poly(N-isopropylmethacrylamide) (PIPMAm) and poly(N,N-diethylacrylamide) (PDEAAm), and some other polymers like poly(vinyl methyl ether) (PVME) exhibit interesting behaviour in aqueous solutions, showing a lower critical solution temperature (LCST). They are soluble at low temperatures but heating above the LCST results in phase separation. On molecular level, both phase separation in solutions and similar collapse transition in crosslinked hydrogels, are assumed to be a macroscopic manifestation of a coil-globule transition often followed by aggregation, as shown for PIPAAm and PDEAAm in water by light scattering and small angle neutron scattering.[1-5] The transition is probably associated with competition between hydrogen bonding and hydrophobic interactions.[6,7]

Their thermosensitivity makes these systems interesting for possible biomedical and technological applications, especially if the polymers are chemically crosslinked (in the form

DOI: 10.1002/masy.200550401

of hydrogels). Stimuli-responsive hydrogels have potential application in the creation of "intelligent" material systems, e.g., as drug delivery systems. A certain similarity to thermal denaturation of proteins in aqueous solutions also makes them interesting from an academic point of view.

Though phase transitions especially in PIPAAm aqueous solutions were extensively studied by various methods, the application of NMR spectroscopy to investigate the phase separation in these systems was rather seldom.[8-14] The present paper provides an overview of our recent ^{1}H NMR studies dealing with temperature-induced phase transitions in D_2O solutions of PVME and PIPMAm, ionized random copolymers of N-isopropylmethacrylamide and sodium methacrylate (MNa), and PIPMAm/PVME mixtures.[15-21] From the methodical point of view, we combined an approach based on measurements of temperature dependences of high-resolution ^{1}H NMR spectra (mainly integrated intensities) with measurements of ^{1}H NMR relaxation times.

Coil-globule Transition and its Manifestation in ^{1}H NMR Spectra

High-resolution ^{1}H NMR spectra of PVME/D_2O solution (c = 4 wt %; PVME was purchased from Aldrich; molecular weight determined by SEC in THF: M_w = 60 500g/mol, $M_w/M_n \cong 3$; tacticity by ^{1}H NMR: 59% of isotactic diads) measured at two slightly different temperatures (307 and 308.5 K) and under identical instrument conditions are shown in Figure 1a. The assignment of resonances to various types of protons of PVME and to residual water (HDO) is shown in spectrum measured at 307 K. While an ordinary spectrum of the polymer in solution was recorded at 307 K, the most significant effect observed at slightly higher temperature but just above the LCST (308.5 K) is a marked decrease in the integrated intensity of all PVME lines. This is evidently due to the fact that at temperatures above the LCST the mobility of most PVME units is reduced to such an extent that corresponding lines become too broad to be detected in high-resolution spectra. All studied PVME/D_2O solutions exhibit a milk-white opalescence at 308.5 K and higher temperatures, so corroborating that a marked line broadening of a major part of PVME units is due to phase separation and formation of compact globular-like structures.

For PVME aqueous solutions the LCST is around 308 K, i.e., well above the temperature of the glass transition of PVME in bulk where values in the range T_g = 191-251 K are

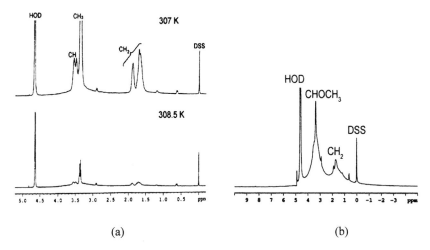

Figure 1. 500.1 MHz ^{1}H NMR spectra of PVME/D$_2$O solutions, c = 4 wt% (a) and 2 wt% (b): (a) spectra at 307 and 308.5 K measured under the same instrumental conditions; (b) spectrum at 312 K measured with spectral width 15 kHz.[16,21]

reported.[22] This means that for this polymer the segmental mobility in globular structures is still relatively high and makes possible to detect directly the broad lines corresponding to PVME units in globular structures by using liquid-state NMR spectrometer. Figure 1b shows ^{1}H NMR spectrum of PVME/D$_2$O solutions (c = 2 wt %) measured at 312 K with larger spectral width and higher amplification. A typical two-component line shape can be seen. Most of PVME units contribute to the broad component (linewidths 408 and 734 Hz for CHOCH$_3$ and CH$_2$ protons, respectively) on which narrow lines of PVME units retaining a high mobility are superimposed. In contrast to PVME, for acrylamide-based polymers (PIPAAm, PDEAAm, PIPMAm) in aqueous solutions the LCST is well below the respective T_g (e.g., for PIPAAm and PIPMAm in bulk, T_g = 403 and 449 K, respectively,[22,23] while LCST for respective aqueous solutions are around 307 and 316 K). Therefore due to much lower mobility of acrylamide-based polymer segments forming globular structures, the corresponding linewidths of the broad component are significantly larger. Linewidths ~ 3.7 and 3.0 kHz were reported for phase-separated segments of PDEAAm and collapsed segments of chemically crosslinked PIPAAm, respectively.[9,13,14]. ^{1}H magic angle spinning NMR

4

spectra and results of MW4 multipulse relaxation measurements ($T_{2\text{eff}}$) have shown that the linewidh 3.7 kHz found for PDEAAm/D_2O solutions and gels is probably associated with isotropic Brownian tumbling of globules as a whole.[13,14]

From a comparison of absolute integrated intensities of NMR lines (measured under conditions allowing to detect only the narrow component, cf. Fig. 1a), the fraction p of phase-separated units (units in globular-like structures) can be determined using the relation[14-17,19,20]

$$p = 1 - (I/I_0) \qquad (1)$$

where I is the integrated intensity of the given polymer line in a partly phase-separated system and I_0 is the integrated intensity of this line if no phase separation occurs. For I_0 we took values based on integrated intensities below the phase transition, using the expected $1/T$ temperature dependence. Figure 2 shows temperature dependences of the fraction p of phase-separated units of PIPMAm as obtained from integrated intensities of CH protons and several concentrations of D_2O solutions[20]; the same dependences were also obtained from analysis of integrated intensities of other proton groups. From Figure 2 it follows that there is no concentration dependence of the transition temperatures in D_2O solutions of neat PIPMAm.

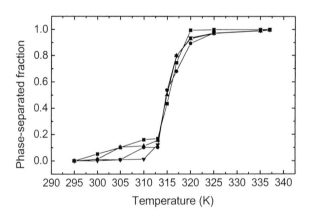

Figure 2. Temperature dependences of the phase-separated fraction p for PIPMAm/D_2O solutions; $c = 0.1$ wt% (■), 1wt% (●), 5 wt% (▲), 10 wt% (▼).

Origin of Nonseparated Portion in PVME/D₂O Solutions

Figure 2 also shows that at temperatures above the phase transition the fraction p of phase-separated segments in PIPMAm/D$_2$O solutions is virtually equal to 1, i.e., virtually no narrow lines were detected in these cases. Similar behaviour was found also for D$_2$O solutions of PIPAAm or PDEAAm.[14,24] For PVME/D$_2$O solutions the fraction of phase-separated PVME segments is $p = 0.85$.[16] As the most probable explanation for 15% of PVME segments that contribute to the minority narrow (mobile) component and do not participate in phase separation we suggested that they probably are from a low-molecular-weight fraction which one can expect for polymer with rather large polydispersity ($M_w/M_n \cong 3$), as in our case. To support this hypothesis, the following experiment was done:[21] PVME/H$_2$O solution was centrifugated for ~ 10 min at 313 K and after seclusion of the phase-separated part the remaining solution was dried and subsequently analyzed by SEC with THF as a mobile phase. The respective SEC chromatogram is shown in Fig. 3. From this figure it follows that really only

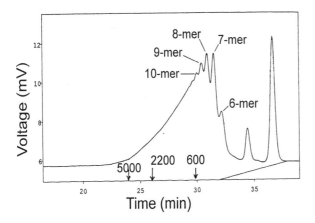

Figure 3. SEC curve (in THF) of the PVME remaining in aqueous solution after seclusion the phase separated part by centrifugation at 313 K and 15 000 rpm (polystyrene standards are shown by arrows).[21]

low-molecular-weight fraction of PVME is present in water solution after removal of the phase-separated polymer. There are several peaks visible on SEC curve, and comparison with polystyrene standards shows that PVME oligomers (6-10 mers) are prevailing in the analyzed sample. One can assume that a certain minimum chain length is a prerequisite for the phase separation as a consequence of the cooperative character of the respective interactions. On the basis of Figure 3, one can assume that for PVME in aqueous solution such minimum chain length amounts on average ~10 monomeric units; however, this figure also shows that some chains with degree of polymerization up to ~50 are not phase-separated at 313 K.

Effects of Ionization

The effect of ionization on the thermotropic phase transition is demonstrated in Figure 4, where the temperature dependences of the phase-separated fraction are shown for ionized copolymers P(IPMAm/MNa) in D_2O solutions (c = 1 and 0.1 wt%) and various ionic comonomer mole fraction (i = 0-10 mole%).[19] MNa units, that are dissociated in aqueous solution (-$CH_2C(CH_3)(COO^-)$-), introduce negative charges on polymer chains. From Figure 4a

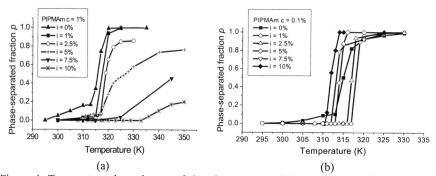

Figure 4. Temperature dependences of the phase-separated fraction p for P(IPMAm/MNa) copolymers in D_2O with polymer concentration 1 wt% (a) and 0.1 wt% (b) and various mole fractions of ionic comonomer i.[19]

(c = 1 wt%) it follows that the transition region shifts towards higher temperatures with increasing concentration of negative charges on the chain and simultaneously the fraction of PIPMAm segments involved in globular structures is reduced. Very similar behavior was found also for c = 10 wt%[19], as well as for hydrogels of chemically crosslinked P(DEAAm/MNa)[13]. In all cases the increasing content of hydrophilic MNa units increases the mobility of polymer segments and higher temperature is necessary to allow hydrophobic interactions to predominate. A completely different behavior was found with dilute solutions ($c = 0.1$ wt%) of P(IPMAm/MNa) copolymers (Fig. 4b). In this case the transition is always rather sharp and interestingly enough, while for $i = 1$ mole% the transition is shifted towards higher temperature in comparison with the neat PIPMAm, further increasing of ionic comonomer mole fraction i gradually shifts the transition towards lower temperatures, even below the dependence for the neat PIPMAm. We assume that this complex behaviour is probably associated with the fact that while for dilute solutions the globular structures are formed mainly by individual macromolecules, for semi-dilute and concentrated solutions intermolecular interactions are important. From Fig. 4 it also follows that the temperature interval and character of the transition can be "tuned" by changing the polymer concentration c and ionic comonomer mole fraction i; this fact might be important for possible practical applications.

Phase Transitions in D_2O Solutions of Polymer Mixtures

^{1}H NMR spectroscopy was used to investigate thermotropic phase transitions in D_2O solutions of PIPMAm/PVME mixtures.[20] In all studied solutions ($c = 0.1 - 5$ wt%) two phase transitions were detected at temperatures roughly corresponding to different LCST of PIPMAm and PVME. While the PVME transition (appearing at lower temperatures) is not affected by the presence of the PIPMAm component in the mixture, the PIPMAAm transition (appearing at higher temperatures) is markedly affected by the presence of the phase-separated PVME component. This is among others demonstrated by the finding that the PIPMAm transition in PIPMAm/PVME mixtures is affected by the concentration of the solution (Fig. 5). While there is no concentration dependence of the transition temperatures for D_2O solutions of neat PIPMAm (cf. Fig. 2), Figure 5 shows that increasing concentration of PIPMAm/PVME mixture leads to the shift of the PIPMAm transition region toward lower temperatures. This shift can be attributed to more frequent PIPMAm-PVME contacts at higher concentrations (some PIPMAm segments might be ketch in PVME globular-like structures). Moreover, for solutions with $c = 5$ wt%, the PIPMAm transition temperatures depend on the composition of PIPMAm/PVME mixture. Similar behavior as described here for

PIPMAm/PVME mixtures was found also for D_2O solutions of PIPMAm/PIPAAm mixtures.[24]

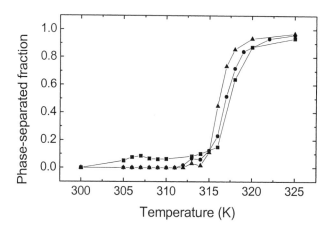

Figure 5. Temperature dependences of the phase-separated fraction p for PIPMAm component in D_2O solutions of mixture PIPMAm/PVME = 1/1 and polymer concentrations c = 0.1 wt% (■), 1 wt% (●), 5 wt% (▲).

In contrast to mixtures of two homopolymers, only single phase transition was found for P(IPMAm/IPAAm) random copolymers by cloud point and ^{1}H NMR measurements.[24,25] At the same time, the phase-transition temperatures strongly depend on the composition of the copolymer. ^{1}H NMR measurements revealed a certain departure from the linear dependence of the transition temperatures on the copolymer composition for a sample with 75 mole% of IPMAm units.[24]

Polymer-water Interactions as Revealed from Measurements of Spin-spin Relaxation

Measurements of ^{1}H spin-spin relaxation time T_2 in PVME/D_2O solutions (with spectra measured in analogous way to that in Figure 1b; T_2 values were measured using the CPMG[26] pulse sequence $90°_x$-$(t_d$-$180°_y$-$t_d)_n$-acquisition) have shown that a very short component ($T_2 <$ 1 ms) dominates the spin-spin relaxation of PVME protons at temperature above the transition

(309.5 K).[18,21] Intensities of the shortest T_2 component do not depend on the concentration of the solution and amount on average ~75%. This value agrees well with the phase-separated fraction $p = 0.85$ as determined from integrated intensities of high-resolution ¹H NMR spectra[16] and confirms that shortest T_2 component corresponds to PVME segments forming globular-like structures. While T_2 values of the shortest component of CHOCH₃ protons virtually do not depend on the polymer concentration, for CH₂ protons the very short component decreases with decreasing concentration of the solution (Figure 6). This shows

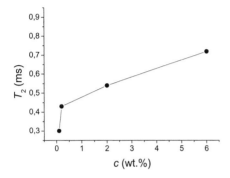

Figure 6. Concentration dependence of a very short component of spin-spin relaxation time T_2 as determined for CH₂ protons in PVME/D₂O solutions at 500.1 MHz and 309.5 K.[18]

that globular-like structures are more compact in dilute solution in comparison with semidilute or concentrated solutions, where globules probably contain a certain amount of water. The result that T_2 values of the short component of CHOCH₃ protons (with dominant contribution of CH₃ protons) do not depend on the concentration of the solution is in accord with results obtained by infrared spectroscopy[27] that at temperatures above the LCST most CH₃ groups of PVME are dehydrated.

To characterize polymer-water interactions in PVME/D₂O solutions, ¹H spin-spin relaxation times T_2 of residual HDO molecules were measured.[18,21] The T_2 values of HDO molecules in solutions of various polymer concentration and measured at temperatures below (305 K) and above (309.5 K) the transition are shown in Table 1. While for dilute solutions ($c = 0.1$ and 0.2 wt%) T_2 values measured at 305 and 309.5 K did not differ too much, a significant difference was found for higher concentrations where T_2 values at 309.5 K were 1 order of

Table 1. ^{1}H spin-spin relaxation times T_2 of HDO molecules in PVME/D$_2$O solutions at 500.1 MHz and two temperatures (305 K and 309.5 K).

c (wt.-%)	T_2 (s)	
	305 K	309.5 K
0.1	4.2	3.1
0.2	5.4	4.3
2	8.7	1.5
6	4.8	0.44
10	4.5	0.74

magnitude shorter than those at 305 K. This shows that in semidilute and concentrated PVME solutions at temperature above the transition there is a portion of HDO molecules that exhibit a lower (spatially restricted) mobility, similarly to the phase-separated PVME. Evidently, this portion corresponds to HDO molecules bound in globular-like structures. The exponential character of T_2 relaxation curves (for HDO) indicates a fast exchange between bound and free sites regarding T_2 values (~1 s); i.e., the lifetime of the bound HDO molecules is < 0.1 s. In such case the observed relaxation time T_{2obs} is given as

$$(T_{2obs})^{-1} = (1 - f)(T_{2F})^{-1} + f(T_{2B})^{-1} \qquad (2)$$

where subscripts F and B correspond to free and bound states, respectively, and f is the fraction of bound HDO molecules. In accord with results of T_2 measurements for PVME protons (Fig. 6), it follows from Table 1 that in dilute solutions the fraction of bound HDO molecules is almost negligible. A certain portion of water molecules bound at elevated temperatures in PVME globular structures in semidilute and concentrated solutions was revealed also from measurements of selective and nonselective spin-lattice relaxation times T_1 of residual HDO molecules.[15,18,21]

From temperature dependences of T_2 of HDO in concentrated D$_2$O solutions of PIPMAm/PVME mixtures, it follows that a certain portion of water is bound both in globular-like structures of PVME and PIPMAm component; a major part of water is bound in globular-like structures of predominant polymer component in the mixture.[20]

Time Dependences of the Spin-spin Relaxation Times T_2: Process of Dehydration

We were interested in knowing whether the amount of water bound in PVME globular structures formed in semidilute and concentrated aqueous solutions is changing with time or not. Figure 7 shows the time dependence of relaxation time T_2 of HDO molecules and time

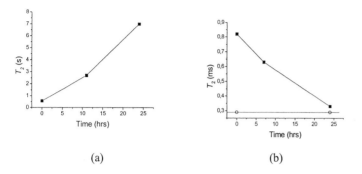

(a)   (b)

Figure 7. Time dependences of spin-spin relaxation times T_2 for HDO protons (a) and a very short component of CH_2 protons of PVME (b) in PVME/D_2O solutions at 500.1 MHz and 309.5 K; $c = 0.1$ wt % (○), 6 wt% (■).[21]

dependences of a very short component of relaxation time T_2 as determined for CH_2 protons of PVME, both measured at 309.5 K.[21] While for $c = 0.1$ wt %, T_2 values of HDO obtained after 24 h only slightly differ from the respective value in Table 1, for $c = 6$ wt% it follows from Figure 7a that T_2 values of HDO very slowly increase with time, reaching after 24 h a similar value as observed at temperature below the transition (cf. Table 1). Simultaneously, for $c = 6$ wt % values of a very short T_2 component of CH_2 protons of PVME (Figure 7b) slowly decrease with time, reaching after 24 h a similar value as found for dilute solution ($c = 0.1$ wt %). Even after very long time ($\sim$ day) we did not observe any sedimentation of the phase-separated part in the studied sample. Therefore, these results evidence that water, originally bound in globular-like structures existing in semidilute and concentrated solutions, is with time very slowly released from these structures. On the contrary, dehydration is rapid in dilute solutions.

T_2 measurements of HDO in D_2O solution ($c = 5$ wt%) of PIPMAm/PVME mixture have shown that dehydration process is about 4 times slower for more rigid globular-like structures of PIPMAm component (at temperature well below the T_g of PIPMAm in bulk) in comparison with globules of the flexible PVME component (at temperatures well above the respective T_g).[21]

Conclusion

Some possibilities of ^{1}H NMR spectroscopy in investigations of structural-dynamic changes and polymer-solvent interactions during the temperature-induced phase transitions in aqueous polymer solutions were shown on examples of several systems studied by us recently. The temperature dependences of the phase-separated fraction p, that can be obtained from integrated intensities in high-resolution ^{1}H NMR spectra, allowed us to quantitatively characterize phase separation in D_2O solutions of negatively charged P(IPMAm/MNa) copolymers or PIPMAm/PVME mixtures. In combination with SEC these measurements have shown that in PVME/D_2O solutions the minority mobile component, which does not take part in the phase transition, mostly consists of low-molecular-weight fraction of PVME. The sponge-like structure containing pores where molecules of water can be accommodated, was suggested for globules existing in semi-dilute and concentrated PVME/D_2O solutions from ^{1}H spin-spin relaxation measurements.[21] With time the bound water is slowly squeezed out by polymer segments, pores disappear, and globules become rather compact. From results obtained on D_2O solutions of PIPMAm/PVME mixtures it follows that the dehydration process is significantly slower for more rigid globular-like structures of PIPMAm in comparison with globules of the flexible PVME. In contrast to this behaviour, in dilute solutions globular-like structures are rather compact and dehydrated already immediately after phase transition (they are mostly formed by single macromolecules).

Acknowledgment

The autor would like to thank his coworkers Lenka Hanyková, Larisa Starovoytova, Michal Ilavský and Petr Holler for their important contributions to the work described here. Support by the Grant Agency of the Academy of Sciences of the Czech Republic (project IAA4050209) is gratefully acknowledged.

[1] S. Fujishige, R. Kubota, I. Ando, *J. Phys. Chem.* **1989**, *93*, 3311.
[2] R. Kubota, S. Fujishige, I. Ando, *J. Phys. Chem.* **1990**, *94*, 5154.
[3] F. W. Zhu, D. H. Napper, *Macromol. Chem. Phys.* **1999**, *200*, 1950.
[4] X. Wang, C. Wu, *Macromolecules* **1999**, *32*, 4299.
[5] J. Pleštil, Y.M. Ostanevich, S. Borbély, J. Stejskal, M. Ilavský, *Polym. Bull.* **1987**, *17*, 465.
[6] H. Maeda, *J. Polym. Sci. Part B: Polym. Phys.* **1994**, *32*, 4299.
[7] Y. Yang, F. Zeng, X. Xie, Z. Tong, X. Liu, *Polym. J.* **2001**, *33*, 399.
[8] H. Ohta, I. Ando, S. Fujishige, K. Kubota, *J. Polym. Sci. Part B: Polym. Phys.* **1991**, *29*, 963.
[9] T. Tokuhiro, T. Amiya, A. Mamada, T. Tanaka, *Macromolecules* **1991**, *24*, 2936.
[10] F. Zeng, Z. Tong, H. Feng, *Polymer* **1997**, *38*, 5539.
[11] M. V. Deskmukh, A. A. Vaidya, M. G. Kulkami, P. R. Rajamohanan, S. Ganapathy, *Polymer* **2000**, *41*, 7951.
[12] A. Durand, D. Hourdet, S. Lafuma, *J. Phys. Chem. B* **2000**, *104*, 9371.
[13] J. Spěváček, D. Geschke, M. Ilavský, *Polymer* **2001**, *42*, 463.
[14] J. Spěváček, L. Hanyková, M. Ilavský, *Macromol. Chem. Phys.* **2001**, *202*, 1122.
[15] J. Spěváček, L. Hanyková, M. Ilavský, *Macromol. Symp.* **2001**, *166*, 231.
[16] L. Hanyková, J. Spěváček, M. Ilavský, *Polymer* **2001**, *42*, 8607.
[17] J. Spěváček, L. Hanyková, L. Starovoytova, M. Ilavský, *Acta Electrotechn. Inform.* **2002**, *2* (3), 36.
[18] J. Spěváček, L. Hanyková, *Macromol. Symp.* **2003**, *203*, 229.
[19] L. Starovoytova, J. Spěváček, L. Hanyková, M. Ilavský, *Macromol. Symp.* **2003**, *203*, 239.
[20] L. Starovoytova, J. Spěváček, L. Hanyková, M. Ilavský, *Polymer* **2004**, *45*, 5905.
[21] J. Spěváček, L. Hanyková, L. Starovoytova, *Macromolecules* **2004**, *37*, 7710.
[22] R. J. Andrews, E. A. Grulke, in: *„Polymer Handbook"*, 4th ed., J. Brandrup, E. H. Immergut, E. A. Grulke, Eds., Wiley, New York 1999, p. VI-201, 215.
[23] M. Salmerón Sánchez, L. Hanyková, M. Ilavský, M. Monleón Pradas, *Polymer* **2004**, *45*, 4087.
[24] L. Starovoytova, J. Spěváček, M. Ilavský, *Polymer*, submitted.
[25] E. Djokpé, W. Vogt, *Macromol. Chem. Phys.* **2001**, *202*, 750.
[26] T. C. Farrar, E. D. Becker, *"Pulse and Fourier Transform NMR"*, Academic Press, New York 1971, p. 27.
[27] Y. Maeda, *Langmuir* **2001**, *17*, 1737.

Macromol. Symp. **2005**, *222*, 15-22

Stereoregular Polyelectrolyte/Surfactant Stoichiometric Complexes

Jean-Michel Guenet

Institut Charles Sadron, CNRS UPR22, 6 rue Boussingault F-67083 Strasbourg Cedex, France
E-mail guenet@ics.u-strasbg.fr

Summary: The molecular structure of polystyrene sulphonate/CTAB stoichiometric complexes has been studied by small angle neutron scattering in solutions and in gels for *atactic* and *isotactic* conformers of the polystyrene moiety. It is found that tacticity has no influence on the molecular structure in solution but plays a role in the gel state.

Keywords: complex; polyelectrolyte; surfactant; tacticity

Introduction

Natural or synthetic polyelectrolytes can be complexed in aqueous solutions by ionic surfactants through electrostatic interaction between the oppositely-charged sites of both constituents [1-4]. Under stoichiometric conditions, the resulting complex is no longer soluble in water but becomes soluble in organic solvents. Curiously enough, these complexes are soluble in alcohol, such as n-butanol or ethanol, so that they cannot be compared to comb-like polymers with short side-chains. In table 1 is shown the behaviour for a few solvents as a function of the dielectric constant ϵ. As can be seen the solubility range seems to be narrow enough as far as ϵ is concerned. Also, some polyelectrolyte-like behaviour is seen at low concentration of alcohol[3], although any dissociation process as occurs for polylectrolytes in water seems hard to accept in ethanol (see figure 1).

Table 1. Solubility of polyelectrolyte/CTAB systems for some solvents vs dielectric constant ϵ.

	ϵ	solubility
decahydronaphthalene	2.2	insoluble
butanol	17.8	soluble
ethanol	25.3	soluble
nitrobenzene	22.9	gels
DMSO	47.2	Phase separates at room temperature
water	80	insoluble

   DOI: 10.1002/masy.200550402

From these alcoholic solutions, films can be cast in which the complex forms ordered structures [5,6] as has been observed with *atactic* polystyrene sulphonate. So far, studies on polystyrene sulphonate have been restricted to the *atactic* variety of this polymer [9]. Here neutron scattering data obtained in the solution state (in *n*-butanol) and the gel state (nitrobenzene) are presented for two different tacticities of the polymer moiety, *atactic* (aPS) and *isotactic* polystyrene (iPS).

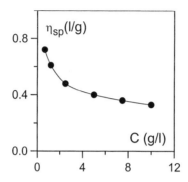

Figure 1. Specific viscosity as a function of concentration for atactic polystyrene sulphonate/CTAB complexes in ethanol. From reference [3].

Experimental

Deuterated styrene was purchased from EURISOTOP and used after proper distillation. Perdeuterated *n-butanol* and perdeuterated nitrobenzene were also purchased from EURISOTOP, and used without further purification. Cetyltrimethylammonium bromide ($CTAB_H$) was obtained from Fluka, hydrogenous *n*-butanol and hydrogenous nitrobenzene were from Aldrich. Perdeuterated atactic polystyrene was synthesized by classical anionic polymerization ($M_w = 10^5$ with $M_w/M_n = 1.26$). Perdeuterated isotactic polystyrene was synthesized by Ziegler-Natta catalysis ($M_w = 1.2 \times 10^5$ with $M_w/M_n = 1.21$). These polymers were sulphonated at the same temperature using concentrated sulfuric acid [10]. In what follows these polymers will be designated as $iPSS_D$ (isotactic) and $aPSS_D$ (atactic). Polyelectrolyte/surfactant complexes are prepared through Antonietti et al.'s method [9].

The small-angle neutron scattering experiments were performed on V4, a small-angle camera located at BENSC (Berlin FRG). A selected wavelength of $\lambda_m = 0.6$ nm was used with a wavelength distribution function of full width at half maximum, FWHD= $\Delta\lambda/\lambda_m \approx 10\%$ (details are available at website http://www.hmi.de/bensc). By moving the sample-detector distance the available q-range was $0.1 < q$ (nm^{-1}) < 3.0 where $q = (4\pi/\lambda_m)\ sin\ (\theta/2)$, θ being the scattering angle.

Samples were prepared directly in quartz cells from HELLMA of optical paths of 1mm (samples in hydrogenous solvents) or 2 mm (samples in deuterated solvents). Details on the neutron scattering data processing are available in reference 11.

Results and discussion

Solutions in n-butanol

Solutions in *n*-butanol have been prepared with concentrations ranging from $C_{comp}= 0.018$g/cm^3 to $C_{comp}= 0.037$g/cm^3. No effect of concentration has been detected. The scattering curves obtained for solutions in *perdeuterated n-butanol* are drawn by means of a Kratky-plot ($q^2I_A(q)$ vs q) in figure 2 for iPSS$_D$/CTA$_H$ systems for aPSS$_D$/CTA$_H$ systems. The important outcome is the absence of effect of the polymer tacticity: the molecular structures of either complexes are identical in the explored q-range and concentration range.

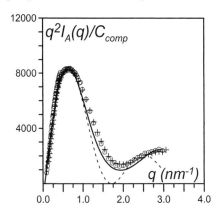

Figure 2. Kratky-plot ($q^2I(q)$ vs q) for solutions in in deuterated n-butanol: iPSS$_D$/CTA$_H$ complexes (+); aPSS$_D$/CTA$_H$ complexes (O); $C_{comp}= 0.034$ g/cm^3.

The scattering curves can be reporduced by using models of rigid cylinders as these systems are expected to take on a bottle-brush structure [12]. The following relation for cylinders of mean length $<L>$ and cross-section radius of gyration r_c [13,14) holds presently:

$$q^2I_A(q) = C_{comp}\mu_L\left[\frac{\pi q <L> -2}{<L>}\right] \times \varphi(qr_c) \tag{1}$$

in which μ_L is the mass per unit length, and where $\varphi(qr_c)$ is a function depending on the cylinder cross-section [11].

For $qr_c \ll 1$ $\varphi(qr_c) \approx 1$ so that μ_L can be determined from the initial slope of the Kratky-plot. One finds $\mu_L = 8500 \pm 2000$ g/nmxmole. Considering the molecular weight of the PSS_D/CTA_H repeating unit (474 g/mole), this figure implies about 18±4 repeating units per nanometre. The value of $<L> \approx 11$ nm is obtained from the intercept q_o with the abscissa ($<L> = 2/\pi q_o$).

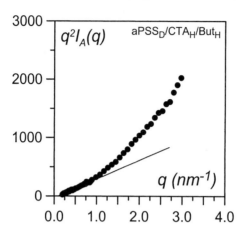

Figure 3. Kratky-plot ($q^2 I(q)$ vs q) for $aPSS_D/CTA_H$ complexes in hydrogenous n-butanol; $C_{comp} = 0.03$ g/cm^3.

The effect of neutron contrast has been examined by dissolving samples of the same polyelectrolyte/surfactant complexes (PSS_D/CTA_H) in *hydrogenous* n-butanol. Again, it must be emphasized again that there is no difference whether one uses $iPSS_D$ or $aPSS_D$. Further support to the cylindrical model is obtained as the data reported in figure 3 by means of a Kratky-representation, while also showing a linear variation at small q, eventually display a strong upturn at large q. In the low-q range, where the variation is linear, the mass per unit length is $\mu_L = 6400 \pm 1000$ g/nmxmole, a value in good agreement with those measured from solutions in perdeuterated n-butanol. The upturn in the large q-range can be accounted for with a cylinder-like structure by introducing a contrast effect in the cross-section scattering. Indeed, the contrast of the polymer moiety and of the CTA wing are of opposite signs ($B_{PSSD}= 12.7$, $B_{CTA}= -0.53$) which can lead to the type of effect (see reference 11 for further details).

The cylinder-like structure arises possibly from the formation of a helical-like structure, especially for the case of iPSS complexes. The existence of helices in solutions has also been reported by Ponomarenko et al. with synthetic polypeptides in dilute chloroform solutions [15]. Note that, as shown by Pringle and Schmidt [14], a helical structure scatters as a cylinder for $q<$

$2\pi/P$, where P is the pitch of the helix, which corresponds to the *zeroth order* layer line.

A fit of the data obtained in perdeuterated *n*-butanol has been attempted in the whole q-range by considering one helical form with the best fit yielding $\gamma_1 = 0.24$ and $r_h = 2.1$ nm. A better fit is obtained by considering the occurrence of two populations of helices with differing cross-section radius as shown in figure 4. The fit yields **helix 1** $r_h = 2.7$ nm/$\gamma_1 = 0.4$ (a), **helix 2** $r_h = 1.75$ nm/$\gamma_1 = 0.33$ (b) with $X = 0.45$ where X is the fraction of helix 1. The existence of two types of helix is suggested by ab-initio calculations performed in the case of isotactic polystyrene, for which it has been shown that *near-tt* arrangements can produce 12_1 and/or 26_1 helices [16,17]. In the case of the 26_1 helix the torsion angles ψ_1 and ψ_2 are of opposite signs which results in the phenyl groups alternating above and beneath the backbone axis. This is reminiscent in some aspect, although to a lesser extent, of the planar zig-zag of syndiotactic polystyrene. In the case of aPS, which consists mainly of syndiotactic tryads, warping of the stablest planar zig-zag form (*tt* conformation) is also likely to give rise to similar helices.

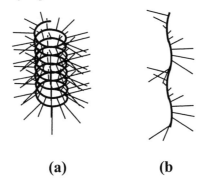

(a) **(b**

Figure 4. Schematic drawings of the two possible helical structures of the polyelectrolyte moiety considered for the fit in figure 2 (solid line).

Gels from nitrobenzene

In nitrobenzene weak gels are produced both for iPSS$_D$/CTA$_H$ and for aPSS$_D$/CTA$_H$ complexes. Typical DSC curves are shown in figure 5. In all cases the melting enthalpies are rather weak ($\Delta H \approx$ 0.2-0.6 J/g for $C_{comp} = 0.25$ g/cm^3) while the overall DSC traces are similar for iPSS$_D$/CTA$_H$ gels and aPSS$_D$/CTA$_H$ gels. The melting temperatures associated with the aPSS$_D$/CTA$_H$ gels are about 20-30°C higher with respect to those from iPSS$_D$/CTA$_H$ gels, which implies that the structures formed from the aPSS$_D$/CTA$_H$ complex are stabler than those formed from the iPSS$_D$/CTA$_H$ complex.

The scattering curve for aPSS$_D$/CTA$_H$/NbH gels differs considerably from that obtained for iPSS$_D$/CTA$_H$/NbH gels which is consistent with the DSC findings (figure 6).

In the low-q range (q< 0.4 nm^{-1}) the scattering curves obtained on iPSS$_D$/CTA$_H$ gels in either isotopes of nitrobenzene can be interpreted by means of a model derived by Guenet for arrays of fibrils of density ρ displaying cross-section polydispersity[18]. The cross-section distribution function is of the type $w(r) \sim r^{-\lambda}$ with two cut-off radii r_{min} and r_{max}. The intensity is written in the so-called *transitional regime* [18]:

$$\frac{q^4 I_A(q)}{C} = 4\pi^2 \rho \left[A(\lambda) q^\lambda - \frac{1}{\lambda r_{max}^\lambda} \right] \Bigg/ \int_{r_{min}}^{r_{max}} w(r) dr \qquad (2)$$

Here, $\lambda \approx 1$ and relation 2 reduces to:

$$\frac{q^4 I_A(q)}{C} = \frac{2\pi^2 \rho}{Log(r_{max}/r_{min})} \times \left[q - \frac{2}{\pi r_{max}} \right] \qquad (3)$$

A linear behaviour is therefore expected whose intercept q_o with the q-axis ($q_o = 2/\pi r_{max}$) allows one to derive r_{max}= 9.2±0.3 nm.

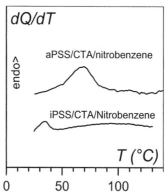

Figure 5. DSC traces obtained at 10°C/m for gels prepared in nitrobenzene: upper curve aPSS/CTA/nitrobenzene gels, lower iPSS/CTA/nitrobenzene gels. C_{comp}= 0.2 g/cm^3.

In principle, the *Porod regime* ($1/q^4$) should be observed after the *transitional regime*. This regime is virtually absent and obliterated by a terminal regime which differs whether the scattering curves have been obtained from gels in hydrogenous nitrobenzene or perdeuterated nitrobenzene (further details in reference 11).

For aPSS$_D$/CTA$_H$/NbH gels there are maxima in a $q^4 I_A(q)$ vs q representation that may suggest a much lower cross-section polydispersity of the fibrils. The low-q domain (q< 0.4 nm^{-1}) can be fitted with a relation for a straight fibril of infinite length and cross-section radius r_c :

$$q^4 I_A(q) = C_{comp}\mu_L\pi q \times \frac{4J_1^2(qr_c)}{r_c^2} \qquad (4)$$

The fit yields r_c= 12.7 nm. That the oscillations are damped at larger q-values may come from a slight polydispersity in cross-section radius.

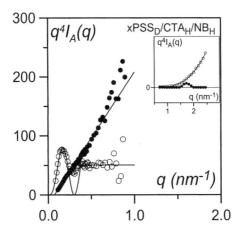

Figure 6. Low-q domain for the intensity scattered by gels in hydrogenous nitrobenzene in a $q^4I(q)$ vs q representation. (O) aPSS$_D$/CTA$_H$, C_{comp}= 0.028 g/cm^3 (•) iPSS$_D$/CTA$_H$, C_{comp}= 0.026 g/cm^3. Solid lines are fits (see text for details). Inset: scattering at larger q which evidences the occurrence of a maximum for aPSS$_D$/CTA$_H$ complexes.

The outcomes from these neutron scattering experiments is consistent with the thermal properties. A lesser cross-section polydispersity together with a larger fibril's cross-section is consistent with a higher melting temperature for aPSS$_D$/CTA$_H$/Nitrobenzene gels with respect to iPSS$_D$/CTA$_H$/Nitrobenzene gels.

Of further interest are the scattering curves in the high-q range in hydrogenous nitrobenzene (NbH) shown in inset of figure 5. The scattering intensity for iPSS$_D$/CTA$_H$/NbH gels increases monotonously while that for aPSS$_D$/CTA$_H$/NbH gels is virtually flat and displays a maximum at q= 1.75 nm^{-1}. This maximum is likely to arise from a lamellar organisation in the fibrils as was reported by Antonietti and coworkers [9]. Conversely, the absence of maximum in iPSS$_D$/CTA$_H$/NbH gels suggests that the helical chains simply pack with a very short-range order.

Concluding remarks

The results highlight the absence of effect of the polymer moiety tacticity on the molecular structure of polystyrene sulphonates/CTAB complexes in solutions. Evidently, the surfactant determines the molecular structure. Conversely, polymer tacticity influences the organization of the gel state. While lamellar structure seems to occur for complexes prepared from atactic polystyrene sulphonate, a much poorer order seems to set in for gels produced from isotactic polystyrene sulphonate.

[1] Hayagawa, K.; Kwak, J.C. *J. Phys. Chem.* **1982** *86* 3866
[2] Goddard, E.D. *Colloids Surf.* **1986** *19* 301
[3] Antonietti, M.; Burger, C.; Thünemann, A. *Trends Polym. Sci.* **1997** *5* 262 and references therein
[4] Ober, C.K.; Wegner, G. *Adv. Mater.* **1997** 9 17 and references therein
[5] Harada, A.; Nozakura, S. *Polym. Bull.* **1984** *11* 175
[6] Taguchi, K.; Yano, S.; Hiratani, K.; Minoura, N.; Okahata, Y. *Macromolecules* **1988** *21* 3336
[7] Ciferri, A. *Macromol. Chem. Phys.* **1994** *195* 457
[8] Thalberg, K. and Lindman, B., *Surfactants in Solution* Vol. 11, Eds. Mittal, K.L. and Shah, D., Pleneum Press, N.Y **1991**.
[9] Antonietti, M.; Conrad, J.; Thünemann, A. *Macromolecules* **1994** *27* 6007
[10] Vink, H. *Makromol. Chem.* **1981** *182* 279
[11] Ray, B.; ElHasri, S.; Guenet, J.M. *Eur. Phys. J. E* **2003** *11* 315
[12] Fredrickson, G.H. *Macromolecules* **1993** *26* 2825
[13] Fournet, G. *Bull. Soc. Franç. Minér. Crist.* **1951** *74* 39
[14] Pringle, O.A.; Schmidt, P.W. *J. Appl. Crystallogr.* **1971** *4* 290
[15] Ponomarenko, E.A.; Tirrell, D.A.; MacKnight, W.J. *Macromolecules* **1996** *29* 8751
[16] Beck, L.; Hägele, P.C. *Colloid Polym. Sci.* **1976** *254* 228
[17] Atkins, E.D.T.; Isaac, D.H.; Keller, A. *J. Polym. Sci. Polym. Phys. Ed.* **1980** *18* 71
[18] Guenet, J.M. *J. Phys. II* **1994** *4* 1077

Rheology and Drug Release Properties of Bioresorbable Hydrogels Prepared from Polylactide/Poly(ethylene glycol) Block Copolymers

Suming Li,* Abdelslam El Ghzaoui, Emilie Dewinck

Centre de Recherche sur les Biopolymères Artificiels, Faculté de Pharmacie, 15 Avenue Charles Flahault, 34060 Montpellier, France
E-mail: lisuming@univ-montp1.fr

Summary: Ring-opening polymerization of L(D)-lactide was realized in the presence of poly(ethylene glycol) (PEG), yielding PLLA/PEG and PDLA/PEG block copolymers. Bioresorbable hydrogels were prepared from aqueous solutions containing both copolymers due to interactions and stereocomplexation between PLLA and PDLA blocks. The rheological properties of the hydrogels were investigated under various conditions by changing copolymer concentration, temperature, time and frequency. The hydrogels constitute a dynamic and evolutive system because of continuous formation/destruction of crosslinks and degradation. Drug release studies were performed on hydrogel systems containing bovine serum albumin (BSA). The release profiles appear almost constant with little burst effect. The release rate depends not only on gelation conditions such as time and temperature, but also on factors such as drug load, as well as molar mass and concentration of the copolymers.

Keywords: bioresorbable; bovine serum albumin; hydrogel; poly(ethylene glycol); polylactide; rheology; stereocomplex

Introduction

Aliphatic polyesters such as poly(lactic acid) (PLA), poly(glycolic acid) (PGA) and poly(ε-caprolactone) (PCL) have been investigated worldwide as biomaterials due to their biocompatibility and degradability.[1-3] These polymers are very atractive biomaterials for temporary therapeutic applications such as sutures, osteosynthetic devices, sustained drug delivery systems (DDS), and scaffolds in tissue engineering.

Hydrogels present growing interest for applications as DDS because of their excellent biocompatibility due to the presence of large amounts of water.[4-6] Bioactive molecules can be physically entrapped in a hydrogel or chemically attached to the polymeric network. Hydrogels are usually formed by a hydrophilic polymer matrix crosslinked chemically through covalent bonds or physically through hydrogen bonds, crystillized domains or

DOI: 10.1002/masy.200550403

hydrophobic interactions. They are particularly interesting for the release of poorly soluble drugs, proteins, genes or nucleic acids as the drugs can be protected from hostile environments, e.g. the presence of enzymes, cells or low pH in the stomach.[7]

Among the various gel systems, degradable, injectable and thermosensitive hydrogels appear the most promising. Kim et al. reported hydrogels prepared from triblock copolymers containing both poly(ethylene glycol) (PEG) and poly(lactide-co-glycolide) (PLGA) blocks.[8-10] A sol-gel transition was observed which depends on the concentration and composition of the copolymers. Hennink et al. prepared a self-assembled hydrogel from enantiomeric PLA oligomers grafted to dextran. The hydrogel was formed due to the stereocomplexation of poly(L-lactide) (PLLA) and poly(D-lactide) (PDLA) blocks.[11-13] Later on, Grijpma and Feijen reported formation of a hydrogel by stereocomplexation of water-soluble PLLA-PEG-PLLA and PDLA-PEG-PDLA copolymers.[14]

In a series of articles, we reported the synthesis, characterization and stereocomplex-induced gelation of PLLA/PEG and PDLA/PEG copolymers.[15-17] The copolymers were synthesized by ring opening polymerization of L(D)-lactide in the presence of mono- or dihydroxyl PEG, using zinc metal as catalyst. Hydrogels were obtained from aqueous solutions containing both PLLA/PEG and PDLA/PEG block copolymers due to stereocomplexation occurring between PLLA and PDLA blocks. In this paper, we wish to report on the rheological properties and drug release behaviors of these bioresorbable hydrogels.

Experimental Section

L-lactide and D-lactide were obtained from Purac and recrystallized from acetone. Dihydroxyl PEG with molar masses of 10000, 12000 and 20000g/mol and monomethoxy poly(ethylene glycol) (mPEG) with molar mass of 5000g/mol were supplied by Fluka. Zinc powder was purchased from Merck. Bovine serum albumin (BSA, Fraction V, pH 7.0) was supplied by Across Organics. Copolymers were synthesized by ring opening polymerization of L(D)-lactide in the presence of mono- or dihydroxyl PEG by using zinc metal as catalyst, as previously reported.[15]

Predetermined amounts of PLLA/PEG and PDLA/PEG copolymers were mixed in 2 ml of distilled water. The aqueous solutions were then centrifuged to yield a homogeneous fluid. Gelation was then allowed to proceed at predetermined temperatures for various periods of time. BSA-containing hydrogels were prepared under similar conditions, BSA being mixed in the aqueous solution before gelation.

2 ml of BSA-containing hydrogel samples were immersed at 37°C in 4 ml of phosphate buffered saline (PBS). The release was regularly monitored by U.V. at 277 nm, using calibration curves obtained from standard solutions. 2 ml of buffer solution were taken out from the release medium, and placed in a 20 mm x 10 mm square quartz cell for each measurement. 2 ml of new PBS solution were then added to the release medium in order to maintain the same volume.

^{1}H Nuclear magnetic resonance (NMR) spectra were recorded at room temperature with a Bruker spectrometer operating at 250 MHz by using DMSO-d_6 as solvent. Chemical shifts (δ) were given in ppm using tetramethylsilane as an internal reference.

Rheological properties were determined on a Carri-Med CSL2 Rheometer of TA Instruments. For all the experiments, a cone-plate measuring geometry was used (steel, 4 cm diameter with an angle of 2 degrees, gap 56 μm). A solvent trap was used to prevent water evaporation. Measurements were realized in the linear viscoelastic range.

The release of BSA was monitored by a Lambda 15 Perkin Elmer UV-Vis spectrophotometer. Circular dichroism (CD) spectra were registered with a Jobin Yvon CD6 instrument. The double monochromator and cell compartment were flushed with nitrogen at a flow rate of 9 l/min. A quartz cell with l = 0.1 cm was used. The CD spectra were obtained from five scans with an integration time of 0.3 seconds.

Results and Discussion

PLA-PEG-PLA triblock copolymers were synthesized by ring opening polymerization of L(D)-lactide in the presence of PEG with molar masses of 10000, 12000 or 20000. Similarly, PLLA-PEG and PDLA-PEG diblock copolymers were synthesized by using mPEG5000 as initiator. Non-toxic Zn powder was used as catalyst instead of stannous octoate which can be more or less cytotoxic.[18,19] Table 1 presents the molecular characteristics of the triblock and diblock copolymers used in this work. The molar ratio of ethylene oxide/lactyl (EO/LA) repeating units was in the range of 4 to 11 for the water solubility of the copolymers.

Table 1. PLA/PEG block copolymers obtained by ring opening polymerization of L(D)-lactide in the presence of PEG or mPEG.

Acronym	Structure	Initiator	Monomer	EO/L A [a]	$\overline{DP}_{PEG}$ [b]	$\overline{DP}_{PLA}$ [c]	$\overline{M}_n$ [d]
1L	$L_{19}EO_{227}L_{19}$	PEG10000	L-lactide	6.1	227	38	12700
1D	$D_{20}EO_{227}D_{20}$	PEG10000	D-lactide	5.6	227	40	13700
2L	$L_{20}EO_{273}L_{20}$	PEG12000	L-lactide	6.8	273	40	14900
2D	$D_{19}EO_{273}D_{19}$	PEG12000	D-lactide	7.3	273	38	14700
3L	$L_{21}EO_{454}L_{21}$	PEG20000	L-lactide	11.0	454	42	23000
3D	$D_{22}EO_{454}D_{22}$	PEG20000	D-lactide	10.5	454	44	23100
4L	$L_{28}EO_{113}$	mPEG5000	L-lactide	4.1	113	28	7000
4D	$D_{27}EO_{113}$	mPEG5000	D-lactide	4.2	113	27	6900

a) calculated from the integration of NMR bands belonging to PEG blocks at 3.6 ppm and to PLA blocks at 5.19 ppm.

b) $\overline{DP}_{PEO} = \overline{M}_{n\,PEG} / 44$

c) $\overline{DP}_{PLA} = \overline{DP}_{PEO} / (EO/LA)$

$\overline{M}_n = \overline{M}_{n\,PEG} + \overline{DP}_{PLA} \bullet 72$

Stereocomplexation is a well known phenomenon for optically active PLA stereocopolymers.[20-24] Stereocomplex can be obtained from co-precipitation of PLLA and PDLA in solution,[20,21] or through cooling from a melt of both polymers.[22] In the case of enantiomeric PLA-PEG-PLA copolymers, stereocomplex was obtained by co-precipitation or solution casting from homogeneous solutions.[23,24]

The phenomenon of stereocomplexation between PLLA and PDLA blocks was used to prepare hydrogels from aqueous solutions containing equal amounts of PLLA/PEG and PDLA/PEG copolymers. Figure 1 shows the evolution of storage modulus (G') and loss modulus (G") of a 14% 1L/1D solution as a function of time at 25°C and at a frequency of 1 Hz. Initially, G" was higher than G', the solution behaving as a viscoelastic solution. Both G' and G" slightly decreased at first and remained constant during the first 60 min. Beyond, the moduli increased continuously, G' increasing faster than G". A crossover point was observed at 7 h. After that, G' became higher than G" and a hydrogel was formed.

Figure 2 shows the changes of storage and loss moduli of the 14% 1L/1D sample as a function of frequency at t = 0 and t = 24h. Both moduli increased with increasing frequency. At t = 0, the storage modulus G' was higher than the loss modulus G", and both moduli increased almost linearly with frequency, which is characteristic of a viscoelastic liquid-like state. In contrast, at t = 24h, G' became higher than G", and both moduli tended towards a plateau at high frequency, which can be assigned to formation of a tridimensional network.

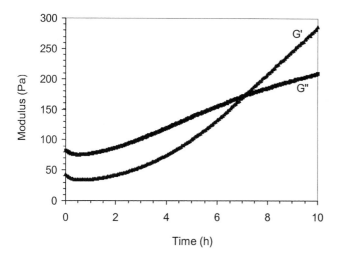

Figure 1. Time-dependent changes of storage modulus (G') and viscous modulus (G") of a 14% 1L/1D sample at 25°C and at 1Hz.

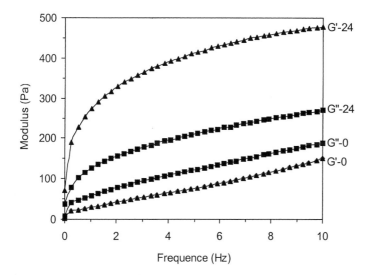

Figure 2. Changes of storage modulus (G') and loss modulus (G") of a 14% 1L/1D solution as a function of frequency at t = 0 and t = 24h at 25°C.

Rheological property changes of the 14% 1L/1D sample were followed for longer periods of time up to 336 h. Figure 3 presents the evolution of G' at 25°C as a function of frequency at different time intervals. The G' value was initially rather low. After 24 h, a large increase was observed for the whole frequency range. At 1 Hz, for example, G' increased from initial 29 Pa to 274 Pa at t = 24 h. Afterwards, G' continued to increase but at a slower rate. After 336 h, G' at 1 Hz reached 507 Pa. These findings show that the hydrogel became more and more consistent as a function of time. A constant increase was also observed for G''. At 1 Hz, G'' increased from initial 58 Pa to 126 Pa at t = 24 h, and to 268 Pa after 336 h. From 24 h to 336 h, G'' remained lower than G', the system behaving as a hydrogel.

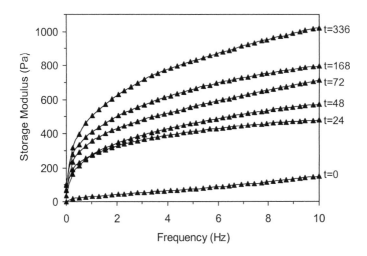

Figure 3. Storage modulus (G') changes of a 14% 1L/1D sample as a function of frequency at t = 0, 24, 48, 72, 168 and 336 h at 25°C.

Hydrogel formation depends also on the molar masses of the copolymers. In the case of PEG10000-derived 1L/1D, hydrogels were obtained at a concentration of 14%, while for PEG20000-derived 3L/3D samples, hydrogels were formed at a concentration of 8%. The higher the molar masse, the lower the concentration at which hydrogels can be obtained.

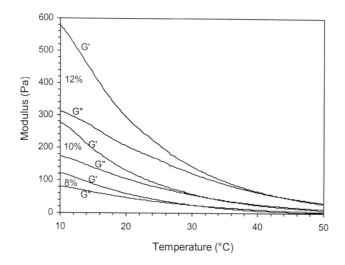

Figure 4. Temperature-dependent modulus changes of 2L/2D samples with concentrations of 8%, 10% and 12% at 1 Hz.

Modulus changes as a function of temperature were followed for 3L/3D samples of 8%, 10% and 12%, as shown in Figure 4. The temperature was increased from 10°C to 50°C at a heating rate of 1°C/min. The modulus values appeared higher for higher concentrations. Both G' and G" decreased with increasing temperature, G' decreasing faster than G". The modulus decrease could be assigned to the destruction of crosslinks at higher temperatures. A gel-sol transition was obtained at 31°C, 33° and 43°C for 8%, 10% and 12% samples, respectively. Therefore, the higher the concentration, the higher the G' and G" values, and the higher the gel-sol transition temperature. The thermo-versibility of the system was examined by following the modulus changes during heating and cooling processes. Data show that the gelation process is reversible although G' and G" values were not exactly the same during cooling and heating processes.

Comparisons were made between 3L/3D samples and 3L ones in order to elucidate the occurrence of L/D interactions and stereocomplexation. Figure 5 shows comparatively the time-dependent modulus changes of 8% 3L/3D and 3L samples. Both moduli of 3L were lower than those of 3L/3D. The 3L sample remained a viscous solution during the 10 h scan, although G' and G" slowly increased. In the case of 3L/3D, it was initially a hydrogel as G' was slightly higher than G". Both moduli increased with time, G' increasing more rapidly than G".

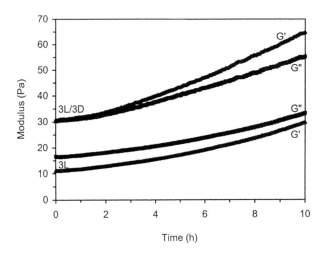

Figure 5. Time-dependent modulus changes of 8% 3L and 3L/3D samples at 25°C and at 1Hz.

In the case of 4L/4D copolymers with a diblock structure and lower molar masses, hydrogels were formed only for concentrations higher than 16%. Figure 6 shows the modulus changes of a 20% 4L/4D sample as a function of frequency at 37°C and at t = 0, 48 and 168 h. The sample was initially in the state of a gel, G' being largely higher than G". After 48 h, both moduli strongly increased, indicating that the gelation process continued at 37°C and the gel became much more consistent. After 168 h, the moduli slightly decreased as compared to values at 48 h, which can be assigned to the partial degradation of the copolymers. In fact, the hydrogel is a dynamic and evolutive system, gelation and degradation occurring simultaneously. Degradation was enhanced at 37°C.

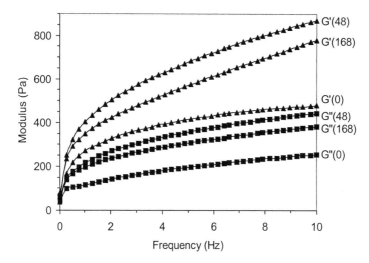

Figure 6. Modulus changes of a 20% 4L/4D sample as a function of frequency at 37°C and at t = 0, 48 and 168 h.

Bovine serum albumine (BSA) was retained as a model drug for release studies. The protein was mixed with the copolymer solution before gel formation. Various BSA-containing hydrogels were prepared under different gelation conditions in order to elucidate the drug release behaviors. Figure 7 shows the BSA release profiles of 20% 2L/2D hydrogels obtained after 90 h at 37°C or after 24 h at 50°C. The release rate appeared almost constant and there was almost no burst release in both cases. On the other hand, BSA release appeared faster from the hydrogel obtained at 37°C than from the one obtained at 50°C, even though the gelation time was longer for the former than for the latter. Nearly 38% and 24% of BSA were released after 360 h, respectively. The difference can be assigned to the fact gelation at 50°C led to much more consistent hydrogel structure than at 37°C, which disfavored drug diffusion.

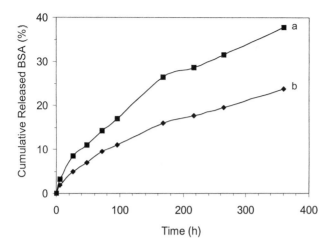

Figure 7. Drug release profiles of 20% 2L/2D hydrogels containing *c.a.* 40 mg of BSA: a) gelation for 90 h at 37°C, b) gelation for 24 h at 50°C.

Circular dichroism (CD) was used to determine whether BSA molecules were denatured after the gelation procedure. Figure 8 shows the CD spectra of original BSA and of BSA released from the hydrogel obtained at 50°C. The two spectra appeared almost identical, indicating that the gelation procedure at 50°C did not denature BSA proteins.

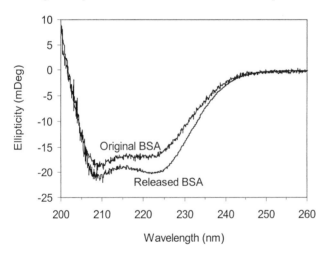

Figure 8. Circular dichroism (CD) spectra of original BSA and of BSA released from the hydrogel obtained at 50°C.

The influence of drug loading on the release behavior was examined by introducing various amounts of BSA in the hydrogel. Figure 9 shows drug release profiles of 20% 2L/2D hydrogels containing *c.a.* 10, 20, 40 and 80 mg of BSA after 24 hours' gelation at 50°C. The release rate decreased for increasing drug amount for gels containing up to 40 mg of BSA. This finding can be assigned to the fact that BSA molecules are molecularly dispersed inside hydrogels with low drug loading and can easily diffuse out, while agglomerates are formed in high drug loading systems. However, the release rate increased when the drug load was further increased to 80 mg, which could be explained by the percolation phenomenon previously described in literature.[25]

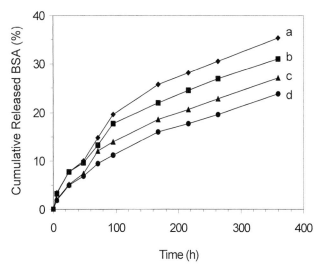

Figure 9. Drug release profiles of 20% 2L/2D hydrogels containing *c.a.* 10 (a), 20 (b), 40 (d) and 80 mg (c) of BSA (gelation for 24 h at 50°C).

Figure 10 shows comparatively the release profiles of 20% 2L/2D and 4L/4D hydrogels containing *c.a.* 40 mg of BSA after 24 hours' gelation at 50°C. It is worth to note that 2L and 2D triblock copolymers possess higher molar masses than 4L and 4D diblock ones. The release rate from 4L/4D gel appeared higher than that from 2L/2D gel, 48% and 24% of BSA being released after 360 h, respectively. The difference can be attributed to the higher molar masses of 2L/2D system which disfavored the diffusion of BSA molecules.

34

The influence of copolymer concentration on the drug release behavior was also examined. Figure 11 presents the release profiles of 4L/4D hydrogels containing *c.a.* 40 mg of BSA after 24 hours' gelation at 50°C, the copolymer concentration of the gel systems varying from 15% to 30%. Similar release curves were observed for different concentrations with almost constant release rate and little burst. Nevertheless, the release rate decreased with increasing copolymer concentration, which can be assigned to the fact that protein diffusion is enhanced with low concentrations.

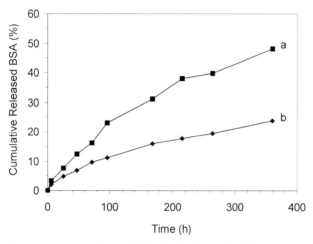

Figure 10. Drug release profiles of 20% 4L/4D (a) and 2L/2D (b) hydrogels containing *c.a.* 40 mg of BSA (gelation for 24h at 50°C).

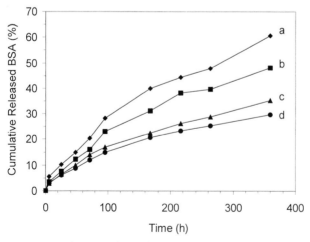

Figure 11. Drug release profiles of 15 (a), 20 (b), 25 (c) and 30% (d) 4L/4D hydrogels containing *c.a.* 40 mg of BSA (gelation for 24 h at 50°C).

Conclusion

Bioresorbable hydrogels were prepared from aqueous solutions containing both PLLA/PEG and PDLA/PEG block copolymers due to interactions and stereocomplexation between PLLA and PDLA blocks. Rheological studies showed that both storage and loss moduli depend not only on the polymer properties such as the molar mass and EO/LA ratio, but also on the factors such as the concentration, temperature, time and frequency. The gelation process is time- and temperature-dependent and the hydrogel is a dynamic and evolutive system because of continuous formation/destruction of crosslinks and degradation. Drug release studies show that the release rate can be adjusted by changing the gelation conditions and factors such as drug load, polymer concentration and molar masses.

[1] S. Li, M. Vert, in *Encyclopedia of Controlled Drug Delivery,* E. Mathiowitz ed., John Wiley & Sons : New York, **1999**, p 71.
[2] R.L. Dunn, in *Biomedical Applications of Synthetic Biodegradable Polymers,* J.O. Hollinger ed. CRC Press: Boca Raton, **1995**, pp 17-31.
[3] S. Li, *J. Biomed. Mater. Res., Appl. Biomater,* **1999**, *48,* 142-153.
[4] Y. Qiu, K. Park, *Advanced Drug Delivery Reviews* **2001**, *53,* 321-339.
[5] A.S. Hoffman, *Advanced Drug Delivery Reviews* **2002**, *54,* 3-12.
[6] J. Heller, R.F. Helwing, R.W. Baker, M.E. Tuttle, *Biomaterials* **1984**, *4,* 262-266.
[7] I. Molina, S. Li, M. Bueno Martinez, M. Vert, *Biomaterials,* **2001**, *22,* 363-369.
[8] B. Jeong, Y.H. Bae, D.S. Lee, S.W. Kim, *Nature* **1997**, *388,* 860-862.
[9] B. Jeong, Y.H. Bae, S.W. Kim, *J. Control. Rel.* **2000**, *63,* 155-163.
[10] B. Jeong, S.W. Kim, Y.H. Bae, *Advanced Drug Delivery Reviews* **2002**, *54,* 37-51.
[11] S.J. de Jong, S.C. De Smedt, M.W.C. Wahls, J. Demeester, J.J. Kettenes-van den Bosch, W.E. Hennink, *Macromolecules* **2000**, *33,* 3680-3686.
[12] S. J. de Jong, B. van Eerdenbrugh, C.F. van Nostrum, J.J. Kettenes-van den Bosch, W.E. Hennink, *J. Control. Rel.* **2001**, *71,* 261-275.
[13] S. J. de Jong, S.C. De Smedt, J. Demeester, C.F. van Nostrum, J.J. Kettenes-van den Bosch, W.E. Hennink, W. E. *J. Control. Rel.* **2001**, *72,* 47-56.
[14] D.W. Grijpma, J. Feijen, *J. Control. Rel.* **2001**, *72,* 247-249.
[15] S. Li, M. Vert, *Macromolecules,* **2003**, *36,* 8008-8014.
[16] S. Li, *Macromol. Biosci.*, **2003**, *3,* 657-661.
[17] A. El Ghzaoui, S. Li, E. Dewinck, M. Vert, Langmuir, submitted.
[18] G. Schwach, J. Coudane, R. Engel, M. Vert, *Polym. Bull.* **1994**, *32,* 617-623.
[19] M.C. Tanzi, P. Verderio, M.G. Lampugnani, M. Resnati, E. Dejana, E. Sturani, *J. Mater Sci.: Mater. Med.* **1994**, *5,* 393-396.
[20] T. Okihara, M. Tsuji, A. Kawagushi, K.I. Katayama, H. Tsuji, S.H. Hyon, Y. Ikada, *J. Macromol. Sci.-Phys.* **1991**, *B30,* 119-140.
[21] H. Tsuji, H.; S.H. Hyon, Y. Ikada, *Macromolecules* **1992**, *25,* 2940-2946.
[22] H. Tsuji, Y. Ikada, *Macromolecules* **1993**, *26,* 6918-6926.
[23] W.M. Stevels, M.J.K. Ankone, P.I. Dijkstra, J. Feijen, *Macromol. Chem. Phys.* **1995**, *11,* 3687-3694.
[24] D.W. Lim, T.G. Park, *J. Appl. Polym. Sci.* **2000**, *75,* 1615-1623.
[25] G. Spenlehauer, M. Vert, J.P. Benoit, F. Chabot, M. Veillard, *J. Control. Rel.*, **1988**, *7,* 217-229.

Gelation Dynamics and Mechanism(s) in Stereoregular Poly(Methyl Methacrylate)s

Alberto Saiani

School of Materials, The University of Manchester, Grosvenor Street, Manchester M1 7HS, United Kingdom
E-mail: a.saiani@manchester.ac.uk

Summary: Chain conformation and gel structure of syndiotactic PMMA thermoreversible gels have been investigated using small angle neutron scattering (SANS). A double helix model for the chain conformation is proposed alongside a gel network model where the fibrils are formed by the proposed double helix and the junctions by the aggregation of 3 double helices. Preliminary results, also obtained by SANS, for stereocomplex gels prepared in bromobenzene are presented.

Keywords: double helix; gelation; small angle neutron scattering; stereocomplex; stereoregular poly(methyl methacrylate)

Introduction

Polymer-solvent interactions have received considerable attention as these interactions, whether desirable or undesirable, are present in most of the applications involving the use of polymers. Of particular interest is the influence of the solvent on polymer conformation, which can strongly influence the resulting properties of the material. For instance, bromobenzene can induce, via polymer-solvent interaction, crystallisation of syndiotactic PMMA but, when the solvent is removed, the sample returns to the amorphous state[1]. It has been noticed that many synthetic (polystyrene, PMMA, poly(vinyl pyridine)) or biologic (agarose, gelatine) polymers exhibit temperature induced conformational transitions in dilute or semi-dilute solutions, which are solvent dependent[2].

A remarkable feature of stereoregular PMMAs is their ability to form helicoidal structures and especially double helix structures, which are often encountered in biological systems but more rarely in synthetic polymers[2]. These double stranded helices can be formed from the bulk, as in the case of isotactic PMMA, for which Kusanagi *et al.* have proposed a symmetric 10_1 double helix for the chain conformation in crystalline samples[3]. They can also be solvent-induced as in syndiotactic PMMA, for which Saiani and Guenet recently suggested the presence of an asymmetric double helix in syndiotactic PMMA gels[4]. A unique feature of stereoregular PMMAs is the existence of the so-called stereocomplex

DOI: 10.1002/masy.200550404

38

that is obtained by mixing isotactic and syndiotactic chains in a 1:2 ratio[5]. Shomaker and Challa have explained this ratio by the existence of an asymmetric double helix where the inner helix is made by an isotactic chain (9_1 helix) around which a syndiotactic chain wraps adopting a 18_1 conformation[6]. Interestingly, syndiotactic PMMA seems to be able to adopt several helicoidal conformations.

Another particularity of stereoregular PMMAs is their ability to form thermoreversible or "physical" gels in a large variety of solvents, yet not necessarily in the same solvent. For instance, while strong aggregation occurs in toluene for highly syndiotactic PMMA, this aggregation seems to be low for isotactic PMMA. The stereocomplex also has the ability to form gels in some solvents and particularly in bromobenzene, for which Fazel *et al.* have proposed the presence of a fibrillar network whose junctions consist of the aggregation of three double helices[7]. The same kind of network model was proposed for syndiotactic PMMA[8] and for κ-carrageenan gels[9]. Interestingly, this last biopolymer also forms double helical structures[1,5,10].

A unique advantage of synthetic polymers is the possibility of deuteration, which allows structural investigations via small angle neutron scattering (SANS). Using various labelling techniques it is possible to investigate the chain conformation in polymer gels as well as the gel structure itself[12]. In this article we will present recent results obtained by A. Saiani and co-workers on syndiotactic PMMA gels as well as preliminary results obtain on sterocomplex gels.

Experimental part

The deuterated and the hydrogenated polymers used were synthesized in our laboratory (for more detail see reference [4]). The tacticity of the hydrogenous polymer was determined in deuterated chloroform by means of proton NMR operating at 200 Hz. The following values were found for the triad arrangements:

syndiotactic PMMA:	iso = 2%	hetero = 9%	syndio = 89%
isotactic PMMA:	iso = 92%	hetero = 3%	syndio = 5%

Due to the low quantity of deuterated material obtained in the synthesis no NMR investigation was carried out. As the thermal behavior of a 5% gel prepared from the deuterated material exhibited no significant difference compared with that of a hydrogenous polymer it was accordingly considered that stereoregularity was little altered.

Gel samples were prepared directly in sealable quartz cells from HELLMA of optical path of 1 mm and 5 mm. After sealing from atmosphere appropriate mixtures of the different constituents, the system was heated up close to the solvent boiling point so as to make clear, homogeneous solutions. Gelation was achieved by a quench to 0 °C for a minimum of 24 hours.

The Small-Angle Neutron Scattering (SANS) experiments on syndiotactic PMMA were carried out on PAXE small-angle camera located at the Laboratoire Léon Brillouin (LLB) (CEN Saclay, France) and on LOQ facility located at ISIS (Rutherford-Appleton Laboratory, Didcot, UK). The experiment on the sterocomplexe gels where carried out on D22 small-angle camera located at the Institut Laue-Langevin (ILL) (Grenoble, France). For more detail on the data analysis see reference [4] and [8].

Results and discussion

Syndiotactic PMMA gels

Recently Saiani et al. have used small angle neutron scattering (SANS) to investigate the chain conformation and gel structure of syndiotactic PMMAs gels. By using different labelling techniques they were able to investigate the chain trajectory as a function of temperature as well as the structure of the gel[4,8].

In the first experiment, partially deuterated samples where used to investigate the conformation of the polymer chain in the middle of the gel. For this purpose gel samples containing 6% (v/v) of deuterated chain where prepared, the total concentration of the samples being 35% (v/v). In this case the coherent scattering observed is due to the deuterated chains. The measurements where performed at 3 different temperatures: room temperature, just above the gel melting temperature and close the solvent boiling temperature. In Figure 1 the scattering curves obtained at 25°C, 81°C and 145°C are presented, by mean of a Kratky representation ($q^2 I_A(q)$ vs q^2), for a gel prepared in hydrogenated bromobenzene solvent.

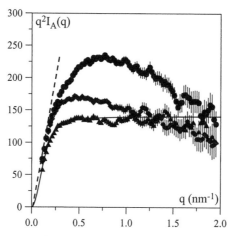

Figure 1. Kratky representation of the intensity scattered by a 35% (v/v) syndiotactic PMMA gel containing 6% (v/v) deuterated chains in bromobenzene: ●, room temperature; ◆, 81°C; ○, 145°C. Solid line stands for the best fit obtained by means of a Debye function, equation (1) (see text and reference [4] for more details).

As can be seen from figure 1 the scattering curves obtained are temperature dependent. At high temperature, close to the boiling point of the solvent, the scattering curve obtained could be fitted using a Debye function:

$$I_A(q) = C_D M_W \frac{2}{q^4 R_G^4} \left[exp(-q^2 R_G^2) + q^2 R_G^2 - 1 \right] \tag{1}$$

where C_D is the deuterated polymer concentration, R_G the mean-square radius of gyration of the polymer and M_W the weight-average molecular weight. In the q range investigated the Debye function is a good approximation for a Gaussian conformation. The values obtained for $R_G \sim 10$ nm and for $M_W \sim 1 \times 10^5$ g mol^{-1} are in good agreement with the results obtained by size exclusion chromatography for the same polymer. This result suggests that at high temperatures bromobenzene is a good solvent for syndiotactic PMMA and the polymer chain adopts a Gaussian conformation.

At lower temperatures it can be seen that the scattering curves changes significantly. In particular at low q the scattering curves present a linear slope in the Kratky representation characteristic of the scattering of rod-like structures. In figure 2 the scattering curve obtained at room temperature is presented with the best fit obtained using a double helix model. It should be noted that single helix models were tested but no satisfactory fit could be obtained.

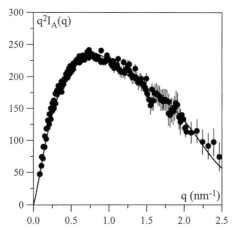

Figure 2. Kratky representation of the intensity scattered by a 35% (v/v) syndiotactic PMMA gel containing 6% (v/v) deuterated chains in bromobenzene at 25°C. The solid line corresponds to the best fit obtained by using adouble helix model, equation (2), and taking into account the presence of 30% Gaussian chains (see text and reference [4] for more details).

As can be seen a good fit of our experimental results is obtained using a double helix model. The scattering by helices has been the subject of a large amount of theoretical work[2]. In the q range used for these experiments helices can be approximated by cylinders. The fit presented in figure 2 has been obtained by considering a double helix model which is schematically depicted in figure 3. For this type of model the scattering can be written:

$$I_A(q) \propto \pi \, C_D \mu_L \left[\frac{f(qr)}{q} + o(q^{-2}) \right]$$ (2)

where μ_L is the linear mass of the double helix and $f(qr)$:

$$f(qr) = \left[2 \, \frac{r_1 J_1(q \, r_1) - \gamma_1 r_1 J_1(q \gamma_1 r_1) + r_2 J_1(q \, r_2) - \gamma_2 r_2 J_1(q \gamma_2 r_2)}{r_1^2(1 - \gamma_1^2) + r_2^2(1 - \gamma_2^2)} \right]^2$$ (3)

where r_1 and r_2 are the outer radii of the inner and outer helix and $\gamma_1 r_1$ and $\gamma_2 r_2$ are the outer radii of the inner and outer helix respectively. The fit presented in Figure 2 was obtained, taking into account the presence of 30% of Gaussian chains, for the following values:

$r_1 \sim 1.1$ nm $\qquad \gamma_1 r_1 \sim 0.6$ nm $\qquad \mu_{L1} \sim 1760$ g mol^{-1} nm^{-1}

$r_2 \sim 2.3$ nm $\qquad \gamma_2 r_2 \sim 1.8$ nm $\qquad \mu_{L2} \sim 2870$ g mol^{-1} nm^{-1}

Figure 3. Schematic representation of the double helix model used to fit the experimental scattering results presented in figure 2.

The values obtained for the radius and the linear mass of the inner helix are in good agreement with the helical model proposed by Kusuyama et al. These authors suggested from X-ray scattering data a single 74_4 helix model for the chain conformation[1]. In our case it was necessary to consider a second outer helix in order to reproduce our results. It has to be noted that in our double helix model there is enough space in the centre of the helix and in between the two strands to accommodate solvent molecules. The existence of polymer-solvent complexes in syndiotactic PMMA gels has been shown using neutron diffraction and NMR. In addition neutron diffraction experiments have been shown to support the proposed double helix model[8].

Just above the melting temperature of the gel it can be seen that the scattering curve changes, suggesting the presence of a different chain conformation in the melt state. As alluded to earlier the linear slope at low q is characteristic of the scattering of rod-like structures, suggesting the presence of a new helical conformation. The probable presence of a significant amount of Gaussian chains makes the fitting of this scattering curve difficult and as yet no satisfactory fit has been obtained.

In order to investigate the structure of the gel itself, in a second series of experiments, syndiotactic PMMA gels were prepared in deuterated solvent in order to label all the chains. In this case the scattering observed is due to the gel network. In figure 4 the scattering curve obtained for a 10% (v/v) gel prepared in deuterated toluene is presented.

The scattering curve obtained could be explained using a network model were the single fibres are formed by the double helix proposed above and the junctions are formed by the aggregation of 3 of these double helices. The position of the scattering peak in figure 4 is very sensitive to the number of double helices in the junctions. Considering more than 3 helices in a junction results in a shift of the peak towards higher q values, while considering 2 double helices in a junction results in the peak being shifted towards lower q values.

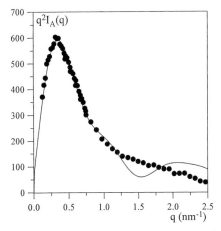

Figure 4. Intensity scattered by a 10% (v/v) syndiotactic PMMA gel prepared in deuterated toluene. The solid line corresponds to the best fit obtained by considering a network model schematically represented in figure 5 (see text and see reference [8] for more details).

Schematic representations of the network structure and the junctions proposed in our model are given in figure 5. It has to be noted that Fazel et al. used the same type of model for stereocomplex gels[7]. For this type of network model and in the explored q range the scattered intensity can be written as follow:

$$q^2 I_A(q) = \pi\, C_P\, \mu_L\, q\, f(q\, r)[1 + 2\, J_0(q\, D)(1 - B)] \qquad (4)$$

where C_p is the polymer concentration, μ_L the linear mass of the double helix, D the distance between the long axes of two adjacent double helices belonging to the same junction and B the weight fraction of junctions. $f(qr)$ is given by equation (3) and the value obtained above for the double helix model were used for r_1, r_2, γ_1 and γ_2.

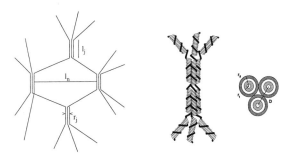

Figure 5. Schematic representation of the gel network model used to fit the experimental scattering results presented in figure 4.

44

The fit presented in figure 4 has been obtained for the following values of D, B and μ_L:

$$D \sim 4.6 \text{ nm} \qquad B \sim 50\% \qquad \mu_L \sim 3400 \text{ g mol}^{-1} \text{ nm}^{-1}$$

As can be seen from figure 5 the value obtained for D is in very good agreement with two double helices being closely packed in a junction. The external radius obtained from our previous experiments for the double helix was ~ 2.3 nm. The linear mass obtained for the double helix is also in very good agreement with the linear masses obtained for the inner and outer strands of the double helix. Finally a 50% weight fraction, which correspond to $\sim 16\%$ volume fraction of junctions is obtained which is a reasonable value for this type of gels.

One of the main limitation when investigating syndiotactic PMMA gels is that the two strands of the double helix can not be labelled independently. Indeed the two strand of the double helix being formed by syndiotactic chains when introducing in the sample deuterated chain they will form randomly inner or outer helices. The model used to fit our SANS results takes in consideration this problem.

Stereocomplex gels

In the case of the stereocomplex gels the double helix is formed by both isotactic and syndiotactic chains. In the model proposed by Shomaker and Challa the isotactic chain adopts a 9_1 single helix conformation and forms the inner strand of the double helix while the syndiotactic chain adopts a 18_1 helix conformation and raps around the isotactic chain to form the outer strand of the double helix[6]. This structure opens the possibility of labelling independently the inner and outer strands of the double helix. Indeed by deuterating the isotactic chain the inner helix will be labelled while by deuterating the syndiotactic chain the outer helix will be labelled.

Preliminary experiments where performed on stereocomplex gels prepared in bromobenzene with a total polymer concentration of 12% (v/v). In order to minimise the number of isolated chain the ratio between isotactic and syndiotactic chains used was 1:2 (4% isotactic chains + 8% syndiotactic chains) in accordance with Shomaker and Challa double helix model. The experiments were performed on D22 which is a high flux spectrometer allowing relatively short acquisition time, typically 5 min, compared to PAXE, typically 2 hours. This allowed us to do real-time experiments.

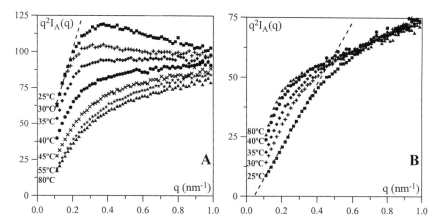

Figure 6. Intensity scattered by a 12% (v/v) stereocomplex gel prepared in bromobenzene. A: sample contains 4% (v/v) of deuterated syndiotactic chain; B: sample contains 4% of deuterated isotactic chains.

The first sample (sample A) was prepared by replacing 4% (v/v) of the protonated syndiotactic chains with their deuterated equivalent. The second sample (sample B) was prepared replacing all the protonated isotactic chains by their deuterated equivalent. In this way the concentration of deuterated polymer for both samples is identical, i.e.: 4% (v/v). The samples were subsequently melted at high temperature, 145°C and then cooled down stepwise (5°C steps) in order to follow the conformational changes of the syndiotactic and isotactic chains during the gelation process. The scattering curves were recorded for each sample at each temperature and are given in figure 6.

Different scattering patterns are obtained for the two samples suggesting, as expected, different chain conformations for syndiotactic and isotactic chains. This seems to be the case even at high temperatures. As the temperature is decreased in both cases the scattering curves change suggesting changes in the conformation of the polymer chains. In the case of the syndiotactic chain no significant changes are observed in the scattering curve from 145°C down to 45°C. In the case of the isotactic, however, no significant changes are observed in the scattering curves down to 35°C, thus suggesting that the syndiotactic chains go through a conformational change 10°C earlier than the isotactic chains. This result implies that the outer strand of the helix form first. Models are currently being developed to confirm this postulation.

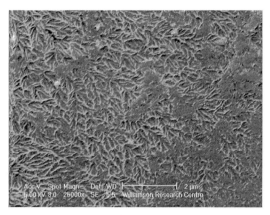

Figure 7. CryoSEM picture of a 12% (v/v) stereocomplex gel prepared in bromobenzene.

Further investigations are necessary in order to elucidate the exact conformation adopted by the two chains and the exact mechanism of the gelation process in relation to the conformational changes of the two polymer chains.

In order to observe the structure of the gel network cryo-electron microscopy experiments were performed. The same gel as used for the neutron scattering experiments were examined. A typical image of the gel structure is presented in figure 7. As can be seen a clear fibrillar structure is present. One characteristic feature of this network seems to be the relatively high number of branches observed. A large fraction of these branches seem to be pendant and not connected to the network.

Conclusion

It has been shown that SANS in combination with labelling techniques is a unique and powerful tool for the investigation of polymer gel structure. Based on our SANS results a double helix model is proposed for the chain conformation is syndiotactic PMMA gel. For the gel structure a fibrillar network model in which the fibrils are formed by the proposed double helix and the junctions by the aggregation of 3 double helices is proposed.

The development of high flux neutron facilities allows real-time investigation of these systems to be performed as has been shown for stereocomplex gels where the conformational changes of the polymer chains could be followed through the gelation process. The results presented here are preliminary results and additional work is in progress in order to elucidate the gelation mechanism of stereoregular PMMAs.

Acknowledgments

The author gratefully acknowledge Dr. J.-M. Guenet from the Institut Charles Sadron, Strasbourg, France, with whom the work on syndiotactic PMMA was performed. The author would like also to gratefully acknowledge Prof. J.S Higgings and Dr. J. Cabral from Imperial College London, UK, in collaboration with whom the work on stereocomplex gels was performed. Finally the author would like to than P. Hill from The University of Manchester for his help with the electron microscope.

[1] H. Kusuyama, N. Miyamoto, Y. Chatani and H., Tadokoro, *Polym. Commun.*, **1983**, 24, 119.

[2] J.-M. Guenet *"Thermoreversible Gelation of Polymers and Biopolymers"*; Acad Press, London, 1992.

[3] H. Kusanagi, H. Tadokoro and Y. Chatani, *Macromolecules*, **1976**, 9, 531.

[4] A. Saiani and J.-M. Guenet, *Macromolecules*, **1997**, 30, 966.

[5] T.G. Fox, B.S. Garrett, W.E. Goode, S. Gratch, J.F. Kincaid, A. Spells and J.D. Stroupe, *J. Am. Chem. Soc.*, **1958**, 80, 1768.

[6] E. Schomaker and G. Challa, *Macromolecules*, **1989**, 22, 3337.

[7] N. Fazel, A. Brûlet and J.-M. Guenet, *Macromolecules*, **1994**, 27, 3836.

[8] A. Saiani and J.-M. Guenet, *Macromolecules*, **1999**, 32, 657.

[9] J.-M. Guenet, C. Rochas and A. Brulet, *Trends in Macromol. Res.*, **1994**, 1, 345.

[10] N.S. Anderson, J.W. Campbell, M.M. Harding, D.A. Rees and J.W.B. Samuel, *J. Mol. Bio.*, **1969**, 45, 85.

[11] K. Buyse, M. Bosco, S. Paoletti and H. Berghmans, *Macromolecules*, **1998**, 31, 9224.

[12] J.S. Higgins and H.C. Benoit, *"Polymer and Neutron Scattering"*, Clarendon Press, Oxford, 1994.

[13] A. Saiani, J. Spevacek and J.-M. Guenet, *Macromolecules*, **1998**, 31.

[14] J. Spevacek and M. Suchoparek *Macromolecules*, **1997**, 30, 2178.

[15] Y. Grohens, P. Carriere, J. Spevacek and J. Schultz, *Polymer*, **1997**, 40, 7033.

Macromol. Symp. **2005**, *222*, 49-63   49

Structure and Properties of Poly(vinyl alcohol) Hydrogels Obtained by Freeze/Thaw Techniques

Rosa Ricciardi, Finizia Auriemma, Claudio De Rosa*

Dipartimento di Chimica, Università di Napoli "Federico II",
Complesso Monte S.Angelo, Via Cintia, 80126 Napoli, Italy
E-mail: derosa@chemistry.unina.it

Summary: The relationships between the structure and the viscoelastic properties of freeze/thaw PVA hydrogels obtained by repeatedly freezing and thawing dilute solutions of PVA in D_2O(11% w/w PVA) in as-prepared and rehydrated states are investigated. Our results indicate that the PVA chains and solvent molecules are organized at different hierarchical length scales, which include the presence of micro- and macro-pores, into a network scaffolding. The porous network is ensured by the presence of crystallites, which act as knots interconnected by portions of PVA chains swollen by the solvent. X-ray diffraction and SANS techniques are used to obtain structural information at short (angstroms) and medium (nanometers) ranges of length scales, concerning the crystallinity, the size of small crystalline aggregates and the average distance between crystallites in PVA hydrogels. Indirect information concerning the structural organization on the large length scales (microns) are provided by viscoelastic measurements. The dynamic shear elastic moduli at low frequency and low strain amplitude, G', are determined and related to the degree of crystallinity. These data indicate that a minimum crystallinity of 1% is required for these PVA samples to exhibit gel behaviour and have allowed obtaining the order of magnitude of the average mesh size in these gels. Finally, it is shown that the negative effect of aging, inducing worse physical and mechanical properties in these systems, may be prevented using a drying/re-hydration protocol able to keep the physical properties of the as-prepared PVA hydrogels.

Keywords: freeze/thaw technique; hydrogels; poly(vinyl alcohol); SANS; X-ray

Introduction

Poly (vinyl alcohol) (PVA) is a semicrystalline synthetic polymer able to form physically crosslinked hydrogels by different methods.[1,2] (PVA) hydrogels prepared by repeatedly freezing and thawing diluted PVA aqueous solutions have attracted much attention in the last years for their many attractive properties as for instance high water content (80-90%wt), dimensional stability at room temperature, high mechanical strength, rubber-like elasticity, lack of toxicity and biocompatibility.[1,3-6]

          DOI: 10.1002/masy.200550405

A limit of PVA hydrogels for practical applications is due to the fact that these gels cannot be stored for a long time, before usage, because their physical and mechanical properties are strongly affected by aging.[7] However, the outstanding physical and mechanical properties of as-formed PVA gels can be preserved even for a long time, drying the sample immediately after the preparation (to avoid aging) and then restored when needed, upon rehydration of the dried samples.[8]

Distinct phenomena may occur during the gelation of PVA water solution through freeze/thaw cycles: phase separation, crystallization and hydrogen bonding. Liquid-liquid phase separation results in a PVA-rich and PVA-poor regions. Then, in the polymer-rich regions, due to the specific properties of PVA, crystallization and hydrogen bonding may take place. As a consequence, the gelation of aqueous PVA solutions through freeze/thaw cycles results in the formation of a porous network in which polymer crystallites act as junction points.[1-3,8-20]

The porous structure of freeze/thaw PVA hydrogels, along with their chemical and mechanical stability, makes freeze/thaw PVA hydrogels attractive matrices in a large variety of biotechnological and biomedical applications.[1,3,21]

The outstanding physical properties of PVA hydrogels derive from their complex structure, where PVA chains and the solvent molecules are organized at different hierarchical scales.

Quantitative information concerning the structural organization of PVA gels formed in mixtures of dimethyl sulfoxide (DMSO) and water on various length scales have been obtained by Kanaya *et al.* using wide[15,17] and small[15-17,19] angle neutron scattering and light scattering[16-18] techniques. These studies confirmed that cross-linking in these gels is ensured by small crystallites.

However, in spite of the numerous studies concerning the use and properties of PVA hydrogels prepared using freeze/thaw techniques, the structure of these systems is still unclear.

In this paper the relationships between the structure and the physical properties of freeze/thaw PVA hydrogels, at different numbers of freeze/thaw cycles, in as-prepared and rehydrated state is analyzed. X-ray diffraction and SANS techniques are used to obtain structural information at short and medium range of length scales, concerning the crystallinity, the size of small crystalline aggregates and the average distance between crystallites in PVA hydrogels and then the results of the structural analysis are related to the viscoelastic properties of freeze/thaw PVA hydrogels.

Experimental Section

Materials. All experiments utilized commercial grade PVA (Aldrich, ref. 36,315-4) with an average molecular weight, $\overline{M}_w$, of about 115000, and a degree of hydrolysis of 98-99%. The ^{13}C NMR spectrum analysis of PVA in deuterated water solution showed that the percentages of *mm*, *mr* and *rr* configurational triads are 22.1, 50.1 and 27.8%, respectively.

PVA hydrogels preparation. Aqueous solutions of PVA of 11%w/w concentration were prepared by dissolving the PVA polymer in deuterated water at 96° C, under reflux and stirring, for about 3 hours. The polymer was entirely dissolved and the obtained homogeneous solutions were slowly cooled to room temperature. We checked that the solutions do not jelly and remain transparent when left at room temperature in a sealed text tube for more then one month.

The freshly prepared PVA solutions were kept for one night, in order to eliminate air bubbles and then poured between glass slides with 1 mm spacers, at room temperature.

Strong physical PVA hydrogel films were obtained by subjecting the polymer aqueous solutions to several repeated freeze/thaw cycles, consisting of a 20h freezing step at –22°C followed by a 4h thawing step at 25°C. In the following sections, the as-formed PVA hydrogels obtained by 1 to 9 freeze/thaw cycles are denoted as GEL-1 to GEL-9 samples.

Dried PVA hydrogel specimens were obtained by keeping in air, at room temperature, the as-formed PVA GEL-*n* immediately after the last *n*-th freeze/thaw cycle. The drying procedure was performed until achieving a constant weight for the PVA hydrogel samples.

Rehydrated PVA hydrogel films were obtained by dipping the so obtained "dried gels" in deuterated water for 1 day (24 hours) or two weeks (14 days).

Gravimetric measurements. Polymer weight concentrations of as-formed and rehydrated PVA hydrogels were determined by weighing each sample in the swollen and in the corresponding dried state.

X-ray measurements. Wide-angle X-ray powder diffraction profiles were collected at room temperature, with a Philips diffractometer using Ni filtered CuKα radiation (λ=1.5418Å) and scans at 0.005 deg(2θ)/s in the 2θ range from 10 to 60°. In order to prevent the sample from drying during the experiment, the diffraction profiles were recorded using a home made brass sample holder placed in a special brass chamber covered with an out-of-focus Mylar film, in an atmosphere saturated with vapors of the mother solution. During the time needed for recording the diffraction patterns ($\approx$ 3h), the weight loss of the sample was less than 2 %wt.

Apparent crystalline dimensions along the $[10\bar{1}]$ lattice direction were calculated by measuring the half-width of the corresponding Bragg reflection and applying the Scherrer formula:[22]

$$t = \frac{k \cdot \lambda}{\beta \cdot \cos\theta} \qquad (1)$$

where t is the apparent crystalline dimension along a given lattice direction, k is a constant ($k=$ 0.89 rad), λ is the wave length of the X-rays, β is the half-width (expressed in radians) and θ is the Bragg angle. Due to the low intensity of the Bragg peak at $2\theta=19.4°$ in the crystalline PVA hydrogels, the standard deviation associated to the so-determined apparent crystalline dimensions is of the order of 3Å.

SANS measurements. Small angle neutron scattering measurements were performed at the KWS2 facility located at the Forschungszentrum of Jülich, Germany. Samples were contained in 1 mm path length quartz cells in order to prevent the drying of gels. Measurements times ranged between 20 min to 8 h. Neutrons with an average wavelength λ of 7 Å and a wavelength spread $\Delta\lambda/\lambda \leq 0.2$ were used. A two-dimensional array detector at three different sample-to-detector distances, 2, 8 and 20 m detected neutrons scattered from the samples. These configurations permitted to collect the scattered intensity in a range of scattering vectors between 0.002 and 0.12 Å^{-1}.

Raw data were corrected for electronic background and empty cell scattering. Detector sensitivity corrections and transformation to absolute scattering cross sections $\left(d\,\sigma(q) \middle/ d\,\Omega \right)$ were made with a secondary Lupolene standard. Raw data were also corrected for intensity of background I_{bck} and intensity of empty cell I_{EC}. Data were then radially averaged and absolute scattering cross sections were obtained.

Shear modulus measurements. Oscillatory dynamic mechanical measurements were performed on PVA hydrogels in as-formed and rehydrated state (24 hours and 14 days). All these rheological measurements were performed using a strain-controlled Rheometrics RFSII rheometer equipped with parallel plates geometry (diameter 25mm).

The experiments were carried out at 25°C. Disc shaped samples (diameter 26mm, thickness ~ 0.5-1mm) of PVA gel were placed between the tools. The samples were protected from drying by a homemade cover which prevented the water from evaporating. This protection ensured the sample stability over a time period long enough (i.e. 1 hour) to perform the measurements of the shear mechanical properties. In all experiments, a weak normal force was applied on the surface of the sample discs in order to avoid the sweeping of the gel from the tool plates. This force ensured a slight compression of the sample. In the frequency sweep experiments, the shear loss (G'') and elastic (G') moduli were measured in the linear viscoelastic regime, for frequencies ranging from 1 to 50 rad/s, at a maximum strain, γ_0, of 0.1-0.6%, depending on the sample. The γ_0 value was determined by preliminary strain sweep experiments, in which the storage and loss modulus were measured as a function of strain at a fixed frequency value of 1 Hz, to check if the deformation imposed to the gel structure by the rheological experiment is entirely reversible.

Each measurement was repeated at least twice, on two different disc specimens from the same sample. The relative error on the storage modulus was of the order of 15%.

Results and Discussion

The polymer concentration in as-formed PVA hydrogels slightly increases with increasing the number of freeze/thaw cycles and, for all gels, is higher than the polymer concentration of the mother solution. It ranges from 12.0 %w/w for GEL-1 to 14.9 %w/w for GEL-9 (see Table 1).

Dried PVA gels, dipped in water, do not dissolve and are able to swell and lead to rehydrated PVA hydrogels with a water content of about 77 - 83 %wt, depending on the number of freeze/thaw cycles (see Table 1).

The X-ray diffraction profile of as-prepared GEL-9 sample along with the X-ray diffraction profile of the dried GEL-9 sample and of the 14 days rehydrated GEL-9 sample are reported in Figure 1 as an example, after subtraction of a straight base line which approximates the background contribution. For comparison, the X-ray diffraction pattern of pure deuterated water, which is the major component of these gels in the as-formed and rehydrated state, is also indicated in Figure 1a,c (dashed line).

The X-ray diffraction profiles of dried PVA hydrogels exhibit the strong diffraction maximum centered around $2\theta = 19.4°$, corresponding to the $10\bar{1}$ reflection of PVA crystals (see Figure 1b, as an example).[8,23]

Table 1. Polymer concentration, total D_2O content, fractional amount of free D_2O, degree of crystallinity (x_c), fraction of swollen amorphous PVA, fraction of crystalline PVA with respect to the sum of the crystalline fraction and the swollen amorphous component (f_c) and apparent crystallite dimensions (along the $[10\bar{1}]$ lattice direction) of freshly prepared PVA/D_2O gels obtained by different numbers of freeze/thaw cycles (fresh) and of gels samples dried immediately after their preparation and dipped in D_2O for 14 days (rehydrated).

gel sample	n° freeze/thaw cycles	gel state	polymer concentration[a] (%w/w)	total D_2O[a] (%w/w)	free D_2O[b] (%)	x_c[b] (%)	swollen amorphous[b] (%)	f_c[b] (%)	Apparent dimensions of crystallites[b] (Å)
GEL-1	1	fresh	12.0	88.0	86.2	0.4	13.4	2.5	28
		rehydrated	17.2	82.8	69.6	1.1	29.0	3.7	39
GEL-3	3	fresh	12.7	87.3	84.2	0.8	15.0	4.8	34
		rehydrated	23.4	76.6	69.8	2.8	27.4	9.2	46
GEL-5	5	fresh	13.4	86.6	80.9	1.1	18.0	5.6	35
GEL-7	7	fresh	14.7	85.3	80.1	1.1	18.8	5.7	38
GEL-8	8	fresh	13.7	86.3	78.5	1.4	20.2	6.3	40
GEL-9	9	fresh	14.9	85.1	78.1	1.4	20.6	6.3	39
		rehydrated	23.3	76.7	72.6	2.9	24.5	10.6	50

[a]Determined by gravimetric measurements. [b]Determined by X-ray powder diffraction experiments.

The diffraction profiles of the as-formed and 14 days rehydrated GEL-9 samples (Figure 1a,c) exhibit two halos centered at $2\theta \approx 28$ and $41°$, as in the diffraction profile of pure water, and a weak peak in the 2θ range 18 - 21° which corresponds to the $10\bar{1}$ reflection of crystalline PVA (Figure 1b).

This result demonstrates the presence of a low amount of small crystalline PVA aggregates in the gel samples in freshly prepared and rehydrated state.

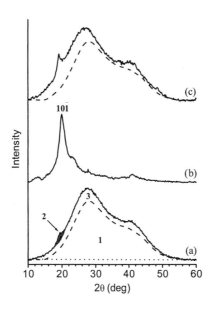

Figure 1. X-ray powder diffraction profiles of freshly prepared PVA hydrogel sample (a), of dried gel (b) and of 14 days rehydrated gel (c), obtained after 9 freeze/thaw cycles (continuous line). The X-ray diffraction profile of liquid D_2O is also reported (dashed line). The crystalline reflection 2 in 2θ range 18 - 21° in (a) is evidenced in grey, whereas the $10\bar{1}$ reflection of the crystalline PVA is indicated in (b).

The X-ray diffraction profile of GEL-n samples (Figure 1a) is considered as the sum of three contributions: a large contribution (area A1) due to the scattering of pure D_2O (dashed curves), a small diffraction component in the range from 18 to 21° due to the crystalline aggregates (area A2) and a third component, (area A3) due to the presence of amorphous PVA swollen by water molecules.

The relative amount of "free D_2O" in the gels, the degree of crystallinity (x_c), the fraction of the swollen amorphous PVA phase and the relative amount of crystalline PVA with respect to the sum of the crystalline and swollen amorphous portions, f_c, may be determined by measuring the areas A1, A2 and A3, as the ratios A1/(A1+A2+A3), A2/(A1+A2+A3), A3/(A1+A2+A3) and A2/(A2+A3), respectively. These values thus obtained and the apparent crystalline dimensions along the [10$\bar{1}$] lattice direction, determined using the Scherrer formula[22], are reported in Table 1.

In freshly prepared gels, the relative amount of "free water" decreases with increasing the number of freeze/thaw cycles (n), whereas in rehydrated samples is nearly constant ($\approx 70\%$). In both kinds of gels the degree of crystallinity increases with increasing n, up to a plateau for n = 5.

Moreover, for freshly prepared gels the apparent crystalline dimensions (along the [10$\bar{1}$] lattice direction) increase with n, ranging from 28 Å for GEL-1 to 40 Å for GEL-9. It is worth noting that, going from GEL-1 to GEL-9, the degree of crystallinity, f_c, increases from 2.5 to 6.3% whereas the polymer concentration slightly increases from 12 to 15 %w/w (see Table 1).

The degrees of crystallinity, f_c, in rehydrated GEL-1, GEL-3 and GEL-9 are 1.1, 2.8 and 2.9 %, respectively; they are slightly higher than the degree of crystallinity of the corresponding as-formed gels (see Table 1). Moreover, the apparent dimensions of crystallites of rehydrated PVA hydrogels slightly increase during rehydration (see Table 1).

Our structural analysis supports previous models proposed in the literature,[2,12] which describe the structure of freeze/thaw PVA hydrogels in terms of a porous polymer network where the crystals act as knots, the polymer segments ensure the connectivity all over the macroscopic gel sample, while free water fills the pores. Water also acts as a swelling agent in the disordered zones of polymer matrix forming hydrogen bonds with the OH groups of PVA chains. The porous walls consist of swollen amorphous PVA. The crystalline knots ensures high dimensional stability and elastic properties of the gel.

The quenching at low temperature, during the freezing step, induces a liquid-liquid phase separation and the formation of ice crystals in polymer-poor phase. The ice crystals, in turn, expel amorphous polymer segments, increasing the polymer concentration in the surrounding environment. The size of ice crystals formed in the polymer-depleted pockets increases through the repeated freeze/thaw cycles. The size of phase-separated domains is of the order of microns, thus accounting for the opaque aspects of PVA hydrogels. Upon thawing, the ice crystals melt and leave the porous structure of the hydrogel unaltered. Practically, water works as a porosigen in PVA solution whereas the polymer network increases its stability during the freezing step.[1,3,24,25]

It is worth noting that, during rehydration of dried PVA gels in D_2O, a certain amount of polymer is released in the solvent. For GEL-1, during the swelling step of the dried PVA hydrogel, the percentage of PVA which is dissolved in D_2O after 2 weeks is of 21.5 %wt whereas for GEL-9, the amount of dissolved PVA does not exceed 4 %wt. This difference in behaviour could be due to the fact that an increase of the number of freeze/thaw cycles improves the stability of the whole gel structure and consequently induces a minor solubility of the PVA chains embedded in the network.

SANS profiles obtained for as-formed PVA GEL-1 and GEL-9 are reported in Figure 2. The SANS profile of the PVA/D_2O (11%w/w) starting solution used for gel preparation is also shown in Figure 2A.

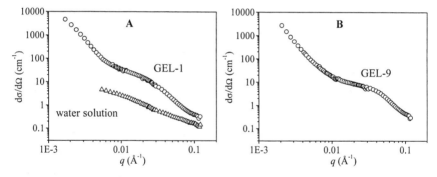

Figure 2. SANS data for (A) as-prepared GEL-1 (○) and PVA solution (△) and (B) as-prepared GEL-9 (○).

The scattering cross section profiles ($d\sigma/d\Omega$) of GEL-n sample present a background higher than the profile of mother solution, mainly due to incoherent scattering contribution. The differences of background are fully accounted for by the different polymer concentration of the

58

various samples (see Table 1).

Inspection of the Figures 2 indicate that SANS data obtained for the PVA starting solution are consistent with the presence of small scattering objects, essentially the individual chains of polymer with a Gaussian coil conformation. Moreover, the scattering cross-section profile of the homogeneous starting solution appears quite different from those of PVA GEL-n samples confirming that, in the gels, PVA chains and solvent molecules are highly organized. In the SANS profiles of the gels we can distinguish three different regions:

1) A region at low q values ($q < 0.009 \text{Å}^{-1}$). In this region the scattering cross section exhibits an upturn, which is clearly not present in the scattering cross section curve of the homogeneous starting solution. In this zone, the data could reflect a supramolecular organization, which we associate to the presence of two separated phases constituted by polymer-rich and polymer-poor regions.

2) A region at intermediate q values ($0.009 < q < 0.035 \text{Å}^{-1}$). In this region an inflexion point is present which gives an indication of the average distance between the scattering crystallites (given approximately by $\frac{2\pi}{q}$). In all gel samples this inflexion point is at $q \approx 0.03 \text{Å}^{-1}$, indicating a distance between crystallites of the order of 200 Å.

3) A region where $0.035 < q < 0.08 \text{Å}^{-1}$. In this zone, scattering cross sections decrease with a power law $\frac{d\sigma}{d\Omega} \propto q^{-D}$, the value of D depending on the number of cycles. More precisely $D \cong 2$ for GEL-1, $D \cong 3$ for GEL-9. Provided we are looking in this region at the boundary structure between two phases, possibly the crystallites and the swollen amorphous in the polymer rich phase, we may apply the surface fractal concept to the function $\frac{d\sigma}{d\Omega}$. According to this concept, the exponent D is related to the surface fractal dimension d_s in a d-dimensional space, through $D = 2d - d_s$. For example, in 3-dimensional space, d_s ranges from 2 to 3, corresponding to a range of D from 4 to 3. If the boundary were smooth, Porod's law ($\frac{d\sigma}{d\Omega} \propto q^{-4}$) would be observed. For our PVA gels, values of D less than 4 suggest that the boundary is not very clear, or it would not be a boundary in the case of GEL-1.

Oscillatory dynamic mechanical measurements were performed on PVA GEL-n in the fresh[13] and rehydrated (24 hours and 14 days) states as a function of n.

The frequency sweep experiments show that both G' and G" moduli of rehydrated GEL-n do not depend on frequency in the range between 1 and 50 rad/s.[14] Moreover, the storage modulus (G') is always higher than the loss modulus (G"). This result indicates that the rehydrated PVA GEL-n samples behave like a highly elastic gel, in agreement with the existence of a network structure and with the results obtained for as-formed GEL-n samples in similar experiments.[13]

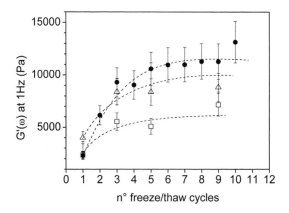

Figure 3. Storage modulus, G', at 1Hz, as a function of the number of freeze/thaw cycles, for PVA hydrogels: as-formed (●); 24h hours rehydrated in D_2O (□); 14 days rehydrated in D_2O (△). Fixed strain amplitude of 0.3-0.6% for rehydrated gel samples and of 0.1% for as-formed samples.[13]

The values of the storage modulus, G', determined at 1 Hz for the 24 hours and 14 days rehydrated PVA GEL-n samples are reported in Figure 3 and compared to those of the corresponding as-formed PVA GEL-n,[13] as a function of the number of freeze/thaw cycles, n. The storage modulus, G', for all rehydrated GEL-n samples, increases with increasing the number of freeze/thaw cycles, tending to a plateau value after the first 3-5 freeze/thaw cycles. The storage modulus, G' for 24 hours rehydrated GEL-n samples is lower than that of freshly prepared samples, except for the GEL-1, for which the storage modulus in the freshly prepared state and after 24 hours rehydration of dried samples are similar. The dried GEL-n samples, at higher number of freeze/thaw cycles, n, need a 14 days rehydration to give G' values comparable to those in the as-prepared state. In the case of GEL-1 sample, 14 days rehydration, instead, results in higher G' values than the G' value in the freshly prepared state.

The drying/ 14 days rehydration procedure gives rise to rehydrated GEL-n samples with G' values comparable to (for n higher than 1) or higher (for n equal to 1) than those in as-formed state.

The relationships between the network structure of the PVA hydrogels and their rheological behavior can be better understood in Figure 4 where the storage modulus values at 1 Hz, G' are plotted as a function of the degree of crystallinity, f_c, determined from X-ray diffraction analysis.

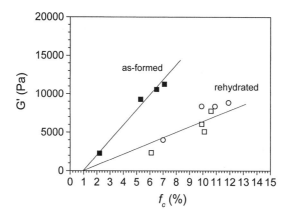

Figure 4. Storage modulus, G', at 1 Hz, as a function of the degree of crystallinity, f_c, for PVA hydrogels in the as-formed state (■), 24 hours rehydrated state (□) and 14 days rehydrated state (○).

For a given PVA concentration, the storage modulus, G', strongly increases as sample crystallinity increases. As a first approximation, the three data sets obtained for the PVA hydrogels (in the fresh, and 24h and 14 days rehydrated states) can be fitted to straight lines. Extrapolation of these lines to G' equal to zero leads to a common value of the sample crystallinity, on the order of 1%. This result indicates that a minimum crystallinity is required for these PVA samples to exhibit gel behavior.

In the hypothesis that our gels may be regarded as a classic network, in spite of the fact that they present a complex hierarchical structure, the storage modulus determined at low strain amplitude and low frequency of our gels (i.e. in the fully elastic, reversible regime), G', may be related to the average number of equivalent units in a "network strand", N, connecting two "ideal" junctions and to the approximate, average value of the network mesh size L_c by use of Equations 2 and 3:[26]

$$G' = \frac{RT}{N_{Av} a^3 N} \phi^{1/3} \tag{2}$$

$$L_c = (\phi)^{-1/3} (C_\infty N)^{1/2} a \tag{3}$$

In Equation 2 N_{Av} is the Avogadro number, G' is the shear modulus of PVA gel samples, R is the gas constant, T the absolute temperature, ϕ the polymer volume fractions of gel in the swollen state, $N_{Av} a^3$ is the molar volume of the solvent, N is the average number of equivalent units with volume equal to the solvent volume (a^3), comprised between two junctions (in a "network strand").[27] For deuterated water $N_{Av} a^3$ is of the order of $18 \mathrm{cm}^3 \mathrm{mol}^{-1}$. In Equation 3 C_∞ is the characteristic ratio of PVA, equal to 8.3.[28]

Values of L_c thus obtained are of the order of 150-200 nm. They indicate that the network mesh size of the gels are ≈ 10 times higher than the average correlation distance between the fringed micelle-like crystallites (≈ 20 nm), determined in SANS analysis.[15,29] Such large values of L_c, indeed, reflect the microscopic heterogeneity of freeze/thaw PVA hydrogels originating from their phase-separated nature, including pores of various dimensions (from nanometers to μm[12,30,31]), highly interconnected through the network scaffolding ensured by the PVA chains.

Conclusions

The structural organization of poly(vinyl alcohol) (PVA) hydrogels obtained by repeatedly freezing and thawing dilute solutions of PVA in D_2O(11% w/w PVA) is investigated by use of X-Ray diffraction and small angle neutron scattering measurements (SANS), as a function of the number of cycles, n. The results of the structural analysis are related to the viscoelastic properties of freeze/thaw PVA hydrogels. The structural analysis techniques and the viscoelastic measurements have been also extended to PVA hydrogels obtained by rehydrating in D_2O the dried samples.

Our results indicate that in freeze/thaw PVA hydrogels the polymer chains and solvent molecules are organized at different hierarchical length scales. The structural organization on the length scales of the order of micron originates from the presence of two separated phases constituted by polymer-rich and polymer-poor regions. The organization on the medium length scales (nanometers) is provided by the presence of small crystallites, highly connected by swollen amorphous tie chains, within the polymer-rich phase. The presence of these tie chains ensures the connectivity of the macroscopic network. The structural organization on the short length scale (angstroms) is essentially provided by the relative arrangement of chains within the crystallites and in the swollen amorphous zones.

It is shown that drying the freeze/thaw hydrogels immediately after their preparation and rehydrating the dried gels for a time long enough ($\sim$ 2 weeks) the structure and properties of PVA hydrogels in the fresh state are restored almost completely avoiding the negative and unpredictable effects of aging.

Acknowledgements

The Centro di Competenza "Nuove Tecnologie per le Attivita` Produttive" Regione Campania P.O.R. 2000-2006 Misura 3.16 is gratefully acknowledged for the financial support.

[1]   Hassan, C.M.; Peppas, N.A., *Advances in Polymer Science*, **2000**, *153*, 37.
[2]   Yokoyama, F.; Masada, I.; Shimamura, K.; Ikawa, T.; Monobe, K., *Colloid and Polymer Science*, **1986**, *264*, 595.
[3]   Lozinsky, V.I., *Russian Chemical Reviews*, **1998**, *67*, 573.
[4]   Urushizaki, F.; Yamaguchi, H.; Nakamura, K.; Numajiri, S.; Sugibayashi, K.; Morimoto, Y., *International Journal of Pharmaceutics*, **1990**, *58*, 135.
[5]   Stauffer, S.R.; Peppas, N.A., *Polymer*, **1992**, *33*, 3932.
[6]   Peppas, N.A.; Scott, J.E., *Journal of Controlled Release*, **1992**, *18*, 95.
[7]   Guenet, J.M., in *Thermoreversible Gelation of Polymers and Biopolymers*, B. R, Editor. 1992: San Diego.
[8]   Ricciardi, R.; Auriemma, F.; De Rosa, C.; Lauprêtre, F., *Macromolecules*, **2004**, *37*, 1921.
[9]   Komatsu, M.; Inoue, T.; Miyasaka, K., *Journal of Polymer Science: Polymer Physics Edition*, **1986**, *24*, 303.
[10]   Watase, M.; Nishinari, K., *Journal of Polymer Science:Part B: Polymer Physics Edition*, **1985**, *23*, 1803.
[11]   Watase, M.; Nishinari, K., *Makromol. Chem.*, **1989**, *190*, 155.
[12]   Willcox, P.J.; Howie, D.W., JR.; Schimdt-Rohr, K.; Hoagland, A.; Gido, S.P.; Pudjijanto, S.; Kleiner, L.W.; Venkatraman, S., *Journal of Polymer Science:Part B: Polymer Physics*, **1999**, *37*, 3438.
[13]   Ricciardi, R.; Gaillet, C.; Ducouret, G.; Lafuma, F.; Lauprêtre, F., *Polymer*, **2003**, *44*, 3375.

[14]   Ricciardi, R.; D'Errico, G.; Auriemma, F.; Ducouret, G.; Tedeschi, A.; De Rosa, C.; Lauprêtre, F.; Lafuma, F., *Chem. Mat. submitted.*

[15]   Kanaya, T.; Ohkura, M.; Kaji, H.; Furusaka, M.; Misawa, M., *Macromolecules*, **1994**, *27*, 5609.

[16]   Kanaya, T.; Ohkura, M.; Takeshita, H.; Kaji, H.; Furusaka, M.; Yamaoka, H.; Wignall, G.D., *Macromolecules*, **1995**, *28*, 3168.

[17]   Kanaya, T.; Takeshita, H.; Nishikoji, Y.; Ohkura, M.; Nishida, K.; Kaji, K., *Supramolecular Science*, **1998**, *5*, 215.

[18]   Takeshita, H.; Kanaya, T.; Nishida, K.; Kaji, K., *Macromolecules*, **1999**, *32*, 7815.

[19]   Takeshita, H.; Kanaya, T.; Nishida, K.; Kaji, K., *Physica B*, **2002**, *311*, 78.

[20]   Takeshita, H.; Kanaya, T.; Nishida, K.; Kaji, K.; Takahashi, T.; Hashimoto, M., *Physical Review E*, **2000**, *61*, 2125.

[21]   Lozinsky, V.I., *Russian Chemical Reviews*, **2002**, *71*, 489.

[22]   Klug, H.P.; Alexander, L.E., in *X-ray diffraction Procedures*, J.W. Sons, Editor. 1959: New York. p. 512.

[23]   Bunn, C.W., *Nature*, **1948**, *161*, 929.

[24]   Chen, J.; Park, H.; Park, K., *J. Biomed. Mater. Res.*, **1999**, *44*, 53.

[25]   Oxley, H.R.; Corkhill, P.H.; Fitton, J.H.; Tighe, B.J., *Biomaterials*, **1993**, *14*, 1065.

[26]   Flory, P.J., *Principles of Polymer Chemistry*, **1953**, *Cornell University Press*, Ithaca NY.

[27]   Rubinstein, M.; Colby, R.H., *Polymer Physics*, **2003**, *Oxford University Press*,

[28]   Brandrup, J.; Immergut, E.H.; Grulke, E.A., *Polymer Handbook*. 1999, fourth edition: John Wiley & Sons, Inc.

[29]   Ricciardi, R.; Mangiapia, G.; Lo Celso, F.; Paduano, L.; Triolo, R.; Auriemma, F.; De Rosa, C.; Lauprêtre, F., *Chem. Mat.* **2005**, *17*, 1183.

[30]   Lozinsky, V.I.; Plieva, F.M., *Enzyme and Microbial Technology*, **1998**, *23*, 227.

[31]   Lozinsky, V.I.; Galaev, I.Y.; Plieva, F.M.; Savina, I.N.; Jungvid, H.; Mattiasson, B., *Trends in Biotechnology*, **2003**, *21*, 445.

Macromol. Symp. **2005**, *222*, 65-71   65

Preparation and Characterisation of Li-Al-glycine Layered Double Hydroxides (LDHs)-Polymer Nanocomposites

Nilwala S. Kottegoda, William Jones*

Department of Chemistry, Lensfield Road, Cambridge, CB2 1EW, UK
Fax: 44 1223 336362; E-mail: wj10@cam.ac.uk

Summary: LiAl-Layered Double Hydroxides, containing glycinate anions, have been prepared using $LiAlO_2$. The glycine containing LDHs were then exfoliated in chloroform. Dispersions of approximately 0.03 g of the LDH in 15 ml of the solvent were possible. Nanocomposites using the exfoliated LiAl-glycine LDH and polyethyleneglycol were prepared by adding appropriate amounts of polymer to the LDH-chloroform dispersion. The clay-polymer nanocomposites were then characterised using powder X-ray diffraction. The thermal and mechanical properties of the composites are reported. The results suggest that composites containing individual (exfoliated) LDH layers were obtained with the mechanical and thermal properties of the composites noticeably superior to those of the parent polymer.

Keywords: dispersion; exfoliation; layered double hydroxides; nanocomposites; polymer

Introduction

Nanocomposites consisting of the assembly of a lamellar nano-sized inorganic host structure (eg silicates) dispersed within a polymer have gained considerable importance [1], Early work on the development of polymer – cationic clay (silicate) nanocomposites (e.g. with montmorillonite and hectorite clays) for materials applications can be traced to the late 1980s and the work on polyamide – clay nanocomposites of the Toyota group [2]. The Toyota nanocomposites were found to possess properties superior to those of conventional composites, primarily, it is believed, because interfacial adhesion between the clay surface and the polymer was maximised. Improved features for the nanocomposites include mechanical properties (such as modulus, strength, heat distortion temperature, thermal expansion co-efficient), enhanced barrier properties, thermal stability, resistance to solvent swelling, flammability resistance, and ablation performances compared to unmodified

          DOI: 10.1002/masy.200550406

polymers and conventional composites.[3-6] It is noteworthy that these improvements were achieved with less than 10% addition of the inorganic material to the overall composite.

From a structural viewpoint, polymer-clay nanocomposites may be broadly classified into two types: (1) intercalated nanocomposites, where either one or a small number of molecular layers of polymer are intercalated into the galleries of the layered material and (2) exfoliated nanocomposites, where the individual layers of the layered compound are dispersed within the polymer matrix[3].

There are three general approaches to the preparation of clay-based nanocomposites. (1) Intercalation of the monomer molecules followed by *in situ* polymerisation; the product may be an intercalated or an exfoliated nanocomposite. (2) Direct intercalation of polymer chains into the host lattice. (3) Transformation of the host material into a colloidal system followed by refoliation in the presence of the polymer.

Layered double hydroxides (LDHs) are another class of layered materials, which are suitable for preparation of nanocomposites.[7] An alternative name frequently used for these materials is anionic clays, reflecting the complementary to the cationic clay family. LDHs present a large variety of compositions and tunable layer charge density [8, 9]. They can easily be synthesised in a relatively pure form under ambient conditions and at low cost.

The use of LDHs in nanocomposite formation has been recently reviewed [7]. Various methods are available for the preparation of LDH-polymer intercalated nanocomposites [3 10-14]. The preparation of exfoliated nanocomposites via transformation of the host lattice into a colloidal system followed by refoliation in the presence of a polymer[15-17] has, however, received less attention due to the difficulties that arise in the exfoliation of LDHs. [18, 19]

In the present contribution, we describe the preparation of LDH/polymer nanocomposites via exfoliation of a Li-Al-glycine LDH in chloroform followed by refoliation in the presence of polyethyleneglycol (PEG). The results from this study suggest that the exfoliated LDH layers are well dispersed within the polymer matrix. In addition, we report a significant improvement in the thermal and mechanical properties of the nanocomposite compared to the parent polymer.

Experimental

Synthesis of LiAl-glycine LDH

0.1 g of LiAlO$_2$ was stirred with 50 ml of a 2 M sodium glycinate solution for 3 days at room temperature under an Ar atmosphere. The resulting slurry was filtered and washed with de-ionised water and dried at room temperature.

Exfoliation of LiAlglycine LDH

0.03 g of the glycine LDH was dispersed in 15 ml chloroform and stirred at room temperature until a clear solution was obtained.

Preparation of LDH-polymer nanocomposite

Appropriate amounts of polyethyleneglycol (MW 10000, white flakes) were added to the LDH – chloroform dispersion (0.03g of LDH in 15 ml of chloroform) and stirred for three days at room temperature. The solvent was then evaporated quickly at 65 °C in order to avoid aggregation of the LDH platelets and the resulting solid was characterised by powder x-ray diffraction (PXRD) and mechanical and thermal properties measured. The composites contained 98, 95 and 90 wt% polymer.

Results and Discussion

PXRD patterns for the parent glycine LDH, the exfoliated and the redeposited product and the nanocomposites prepared with different amounts of clay loading are shown in Figure 1. The first reflection of the glycine LDH suggests an interlayer repeat distance of 12.40 Å. Taking into account that the thickness of the layer is 4.8 Å, a gallery height of 7.6 Å is suggested. An interlayer repeat of 12.40 Å is considerably higher than the value of 8.1 Å reported by Hibino and Jones for a MgAl-glycine LDH [20]. Since the length of the glycine anion is approximately 3.9 Å, the most likely arrangement of anions in the LiAl-glycine LDH is as a bilayer of glycinate anions. The reflections around 11.8 and 23.54 °2θ may be due to a second LDH phase and the corresponding interlayer spacing for this phase is 7.5Å. The crystalline impurities may be unreacted LiAlO$_2$ (ICSD number 38-1464), gibbsite (ICSD number 01-0263) and Li glycine salt. However, it is not possible to unambiguously assign phases to the extra reflections since there are some reflections common for all three impurity phases. None of the impurities were removed by repeated washing with deionised water.

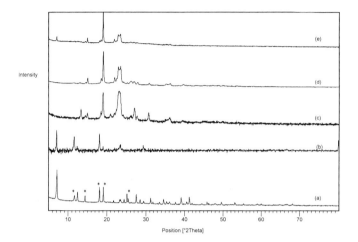

Figure 1. Powder X-Ray patterns for (a) parent LiAlglycine LDH, (b) LDH resulted after exfoliation and redeposition of glycine LDH, (c) polyethyleneglycol, (d) nanocomposite with 5% LDH loading, and (e) nanocomposite with 10 % LDH loading nanocomposite (* Li glycine salt impurities).

The formation of nanocomposites with exfoliated clay materials (with 2% and 5% LDH loading) is suggested principally by the absence of basal reflections associated with the LDH. The absence of the (00l) reflection is, however, not a direct evidence of the formation of an exfoliated LDH–polymer nanocomposite because for a 2% clay loading is likely to under the minimum detection limit for PXRD and at 5% clay will be very close to the minimum. The presence of (00l) reflections in the nanocomposite with 10 % LDH loading (Figure 1(e)) indicates some ordering of the LDH material within the polymer matrix for this material. However, the interlayer spacing of the LDH is similar to that of the original glycine LDH suggesting that the product is not a mixture of polymer intercalated and exfoliated nanocomposites. Instead it is likely that tactoids of the original inorganic material are present within the matrix together with exfoliated layered material.

Figure 2 indicates the thermogravimetric curves of the polymer and the 10 wr% LDH-polymer composite.

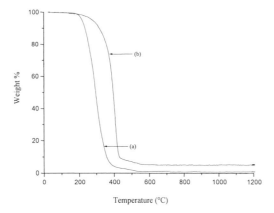

Figure 2. TGA curves of (a) polyethyleneglycol, (b) nanocomposite with 10% LDH loading.

Three weight losses are observed for the parent polymer and the nanocomposite. The first (minor) weight loss (up to *ca* 200 °C), corresponds to elimination of water, the second to (*ca* 400 °C) dehydroxylation and partial decomposition of the polymer and the final step (> 400 °C) to oxidative elimination of the carbonaceous residue derived from the initial polymer degradation[13]. There is a clear increase in the temperature regions for weight loss after addition of the LDH (approximately 100 °C). A factor leading to the increased stability will be the well dispersed inorganic material acting as a gas barrier, preventing evolution of volatile gases from the thermally decomposed products [11].

Table 1 summarises the maximum load and the tensile strength values of the composite materials with 2 and 5wt% loadings.

Table 1. Mechanical properties of the nanocomposites.

Sample	Area of the fracture (mm^2)	Maximum force (N)	Tensile strength (N/mm^2)
Polyethyleneglycol	9.2	0.40	0.04
Nanocomposite with 2% LDH	2.30	0.42	0.18
Nanocomposite with 5% LDH	5.10	0.64	0.12

The data clearly shows that the LDH-polymer nanocomposites are mechanically more stable than the parent polymer. The maximum tensile strength was observed with 2% LDH loading, with a value 45% higher than that of the parent polymer.

Conclusions

We have demonstrated that LiAl LDHs can be readily prepared from $LiAlO_2$ as the starting material. Exfoliated LiAlglycine LDHs were used to prepare LDH-polyethyleneglycol nanocomposites. XRD measurements indicate that the individual LDH platelets are well dispersed within the polymer matrix up to 5% LDH loading. The presence of exfoliated LDH platelets within the polymer matrix improves the mechanical and thermal properties of the parent LDH.

In preparation of nanocomposites it is important to consider the compatibility between the polymer material and the LDH because lack of adhesion between the two compounds may lead to phase separation[11]. The amino groups of the glycinate anion interacts with the hydroxyl groups of the polymer through H-bonding in order to generate adhesion between the LDH and polymer making the exfoliated LDH and the polymer more compatible leading to LDH – polymer with dramatically improved mechanical and thermal properties over those of the parent polymer. TEM/SEM investigations are currently under way as well as FTIR and Raman spectroscopy. The use of other polymer matrices will also be presented.

Acknowledgements

The authors wish to thank Mr. Brian Whitemore for assistance with the mechanical property testing. NSK is grateful for an ORS award and support from the Cambridge Commonwealth Trust.

[1]"Polymer - Clay Nanocomposites", T. J. Pinnavaia, G. W. Beall, Eds., Wiley Series in Polymer Science, New York, 2001.
[2]Y. Fukushima, S. Inagaki, *J. Inclusion Phenom.* **1987**, 5, 473.
[3]Z. Wang, J. Massam, T. J. Pinnavaia, in: "Polymer - Clay Nanocomposites", T. J. Pinnavaia, G. W. Beall, Eds., Wiley Series in Polymer Science, New York, 2001.
[4]A. Usuki, M. Kawasumi, Y. Kojima, A. Okada, T. Karauchi, O. Kamigaito, *J. Mater. Res.* **1993**, 1774.
[5]J. H. Chang, Y. U. An, D. H. Cho, E. P. Giannelis, *Polymer* **2003**, 44, 3715.
[6]C. S. Triantafillidis, P. C. LeBaron, T. J. Pinnavaia, *Chem. Mater.* **2002**, 14, 4088.
[7]F. Leroux, J. P. Besse, *Chem. Mater.* **2001**, 13, 3507.

[8] S. P. Newman, W. Jones, in: "Supramolecular Organisation and Materials Design", W. Jones, C. N. R. Rao, Eds., Cambridge University Press, Cambridge, 2002.

[9] A. D. Roy, C. Forano, J. P. Besse, in: "Layered Double Hydroxides: Present and Future", V. Rives, Eds. Nova Science Publishers, New York, 2001.

[10] E. M. Moujahid, J. Inacio, J. P. Besse, F. Leroux, *Microporous Mesoporous Mater.* **2003**, 57, 37.

[11] H. B. Hsueh, C. Y. Chen, *Polymer* **2003**, 44, 5275.

[12] L. Vieille, C. Taviot-Gueho, J. P. Besse, F. Leroux, *Chem. Mater.* **2003**, 15, 4369.

[13] C. O. Oriakhi, I. V. Farr, M. M. Lerner, *J. Mater. Chem.* **1996**, 6, 103.

[14] T. Challier, R. C. T. Slade, *J. Mater. Chem.* **1994**, 4, 367.

[15] B. Li, Y. Hu, J. Liu, Z. Chen, W. Fan, *Colloid. Polym. Sci.* **2003**, 281, 998.

[16] W. Chen, B. J. Qu, *J. Mater. Chem.* **2004**, 14, 1705.

[17] W. Chen, B. Qu, *Chem. Mater.* **2003**, 15, 3208.

[18] S. O'Leary, D. O'Hare, G. Seeley, *Chem. Commun.* **2002**, 1506.

[19] M. Adachi-Pagano, C. Forano, J. P. Besse, *Chem. Commun.* **2000**, 91.

[20] T. Hibino, W. Jones, *J. Mater. Chem.* **2001**, 11, 1321.

Macromol. Symp. **2005**, *222*, 73-79

Thermoreversible Gelation of Syndiotactic Polystyrene in Naphthalene

Sudip Malik,[1] *Cyrille Rochas,*[2] *Bruno Démé,*[3] *Jean-Michel Guenet**[1]

[1] Institut Charles Sadron, CNRS UPR22, 6 rue Boussingault, 67083 Strasbourg Cedex, France

[2] Laboratoire de Spectrométrie Physique CNRS-UJF UMR5588, 38402 Saint Martin d'Hères Cedex, France

[3] Institut Laue-Langevin, 6 rue Jules Horowitz, BP 156, 38042 Grenoble Cedex, France

Summary: Investigation into syndiotactic polystyrene/naphthalene systems of concentrations ranging from 0 to 78% (w/w) have been carried out by electron microscopy, DSC, and neutron diffraction. It is found that a fibrillar morphology is produced in this solvent, similar to that observed for thermoreversible gels. The temperature-concentration phase diagram suggests the existence of two compounds of differing stoichiometries. Neutron diffraction experiments confirm the existence of compounds in this system.

Keywords: polymer-solvent complex; syndiotactic polystyrene; T-C phase diagram; thermoreversible gel

Introduction

Syndiotactic polystyrene (sPS) has received great interest as an attractive material due to its high crystallization rate, high melting temperature, low specific gravity and good chemical resistance [1]. sPS possesses very complex polymorphic behaviour, which, by simplification, can be described in terms of two crystalline forms, α and β, containing all trans conformation with identity period 5.1Å and two forms, δ and γ, containing helical chain conformation with identity period of 7.7Å [2] sPS has a tendency to form polymer-solvent compounds (also designated sometimes as crystallosolvates, intercalates or chlathrates), which are due to formation of solvent-induced crystalline δ-form, with a large variety of solvents such as benzene [3], toluene [4], chloroform, bromoform [5], decalin etc. by exposing solid polymer in the liquid solvent or vapour (solvent-induced-process) [6] or by cooling homogeneous solution (solution-cast-process) [7]. In previous reports, sPS complexes were prepared from solvents that are liquid at room temperature and also toxic. For the sake of using safer solvents, we have considered here naphthalene that is far less harmful to human body, and which is easier

DOI: 10.1002/masy.200550407

to handle as it is in the solid state in room temperature. Naphthalene possesses well-known sublimation property.

Experimental

Syndiotactic polystyrene (sPS) samples, hydrogenated and deuterated, were synthesized following a method devised by Zambelli and co-workers [8]. The content of syndiotactic triads characterized by ^{13}C-NMR was found to be over 99%. The molecular weight characterization of these samples was performed by GPC in dichlorobenzene at 140°C and yielded the following data: $M_w = 1.0 \times 10^5$ g/mol with $M_w/M_n = 4.4$ for sPSH; $M_w = 4.3 \times 10^4$ g/mol with $M_w/M_n = 3.6$ for sPSD.

Naphthalene was purchased from Aldrich and used without further purification.

The thermal behaviour of the gel was investigated by means of Perkin Elmer DSC 7. Pieces of gel prepared beforehand in test tube were introduced into "volatile sample" pans that were hermetically sealed. The systems were melted so as to obtain a homogeneous solution, and then cooled to 20°C. The thermograms were recorded at a scan rate of 5°C/min. The weight of the sample was checked after each experiment and the instrument was calibrated with indium before each set of experiment.

A film of sPS/naphthalene was dried in vacuum at room temperature and was coated with gold by sputtering technique under argon atmosphere and observed under microscope (Hitachi S-2300) operating voltage ranging from 15 kV to 25 kV.

Neutron diffraction experiments were carried out on D16, a diffractometer located at Institut Laue Langevin, Grenoble, France. It is a two circle diffractometer equipped with a position sensitive ^{3}He multidetector made up with 128x128 wires. The distance between consecutive wires is 2 mm. The diffractometer operates at wavelength $\lambda = 0.454$ nm obtained by diffraction of the neutron beam onto a pyrolytic graphite mosaic-crystal (mosaicity = 0.7°) oriented under Bragg condition. Momentum transfer $q = (4\pi/\lambda)\sin(\theta/2)$ were ranging from $q = 2$ to 12 nm^{-1}. Detector normalization and correction of cell efficiency were achieved with the spectrum given off by water. The samples were prepared directly in quartz tubes of 3 mm inner diameter. After introducing a mixture of polymer (sPSD) and solvent these tubes were sealed from atmosphere. Homogeneous solution were obtained by heating and were then quenched to room temperature.

Results and Discussion

Samples from sPS in naphthalene are solid in the room temperature but become transparent above the melting temperature of pure naphthalene (82°C), and possess all the characteristics of thermoreversible gels. The morphology of this system, as observed by SEM, does reveal a fibrillar network morphology (see fig. 1) similar to that reported for thermoreversible gels [7]. The mesh size is in the micrometer range while the fibrils cross-sections are in the nanometer range (around 20-40 nm).

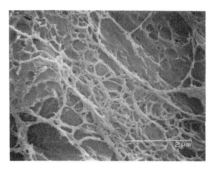

Figure 1. Scanning electron micrograph of 30%(w/w) sPS-naphthalene dries gels.

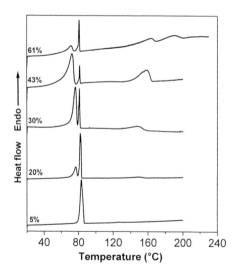

Figure 2. The representative DSC traces of sPS-naphthalene system at indicated concentration.

Typical differential scanning calorimetry (DSC) thermograms of sPS in naphthalene at indicated polymer concentration are presented in Figure 2. As can be seen there are two domains, a low temperature region and a high temperature region. The temperature associated with the first domain is due to the melting of the free naphthalene present in the system. It is interesting to note that the peaks due to naphthalene melting at the temperature range 82 – 68°C are bifurcated at polymer concentration (C_{pol}) 20-66% (g/g). The reason of bifurcation of the solvent peak is probably due to the fact that increasing the polymer concentration produces smaller and smaller free solvent domains with a melting point lower than that of infinite crystals .

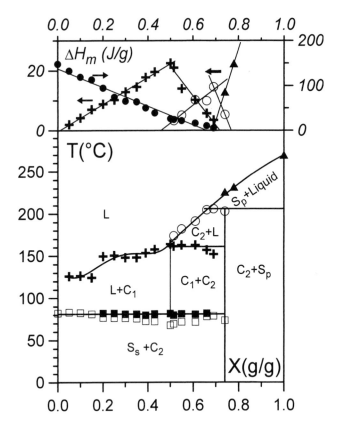

Figure 3. Temperature-concentration phase diagram and Tamman's diagram for sPS-naphthalene systems.

In the high temperature domain, one melting endotherm (low-melting endotherm) is observed up to a concentration $C_P = 50\%$ (w/w) while for the high concentrated systems a second melting endotherm (high-melting endotherm) appears at higher temperature. The temperature associated with low-melting endotherm, initially , remains constant up to $C_P = 15\%$, after this it jumps to 149°C at $C_P = 20\%$, and then increases with increasing concentration. The temperature associated with high-melting endotherm increases continuously with increasing concentration of sPS investigated here. The temperature- concentration phase diagram drawn in figure 3 presents these different types of behaviour. The variations of the enthalpies associated with each endotherm as function of polymer concentration (Tamman's diagram) are presented in the same figure.

The enthalpy associated with the melting of naphthalene is gradually decreasing with increasing polymer concentration and becomes zero at $C_P = 70\%$. The enthalpy associated with low melting endotherm at high temperature region increases linearly up to a polymer concentration $C_P = 50\%$ and then decreases to become zero at about $C_P = 74\%$. On the other hand, the enthalpy associated with high melting endotherm increases continuously. The enthalpy variations together with the shape of phase diagram are consistent with the existence of the two compounds C_1 and C_2 of different stoichiometries. C_1 is a singularly melting compound that transforms to C_2 at low temperature. The stoichiometry of the compound C_1 is determined by the maximum of the enthalpy associated with T_{low}, namely about 0.9 naphthalene molecules per monomer unit. The stoichiometry of compound C_2 is given by the concentration at which the enthalpy of low melting endotherm goes to zero ($C_P = 74\%$) , namely one naphthalene molecule per four monomeric unit, and is incongruently-melting compound.

Neutron diffraction is an appropriate tool for studying polymer-solvent compound as deuterium-labelling of either component allows one to get four structure factors without significant change in the molecular arrangement. The existence of polymer-solvent compounds can be therefore confirmed from neutron diffraction as is illustrated by the general expression for the intensity diffracted by a binary system composed of polymer(p) and solvent(s) [9,10]:

$$I(q) = \overline{A}_p^2(q)S_p(q) + \overline{A}_s^2(q)S_s(q) + 2A_p(q)A_s(q)S_{ps}(q) \qquad (1)$$

where $A(q)$ and $S(q)$ with appropriate subscripts are the coherent scattering amplitude and the structure factor of the polymer and of the solvent, and $S_{ps}(q)$ is a cross-term between polymer and solvent. This term is only meaningful when polymer-solvent compound are dealt with, which results in an alteration of the ratio of diffracted intensities with respect to one another when different labelled species are used. Conversely, if no complex is formed, this cross term vanishes and the diffraction pattern of the crystallized polymer is independent of labelling. Therefore, if significant variation of this ratio is observed when replacing one species by its labelled counterpart then polymer- solvent compounds are present beyond doubt.

Here we shall discuss only the qualitative aspect of the diffraction patterns that are presented in figure 4 and figure 5 for $C_P = 20$ wt % and $C_P = 73$ wt % respectively. As can be seen in figure 4 , the intensity ratio does vary with the solvent labelling. For instant, the reflection at 7.3 nm^{-1} (indicated by arrow) is much stronger for sPSD/napD (curve a) than that for sPSD/napH (curve b) while that at 5.4 nm^{-1} varies little. This clearly supports the existence of a sPS/Naphthalene compounds. Note that the strong reflection at 8.7 nm^{-1} arises from the free, crystallized naphthalene and it vanishes when the system is heated above 80°C (curve c).

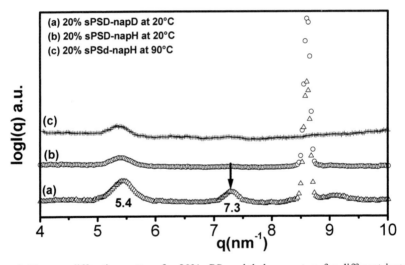

Figure 4. Neutron diffraction pattern for 20% sPS-naphthalene system for different isotopic solvent labelling (as indicated).

In figure 5 the relative intensities of the peaks at $q = 5.4$ nm^{-1} and $q = 7.3$ nm^{-1} are significantly altered with changing solvent labelling. This indicates clearly the presence of another complex at $C_p = 74\%$ (g/g) in the sPS/naphthalene system. This indication supports the phase diagram as well as the existence of two polymer-solvent compound in this system.

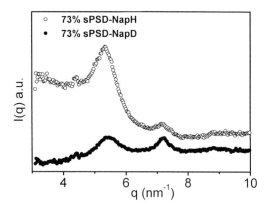

Figure 5. Neutron diffraction pattern for 74% sPS-naphthalene system for different isotopic solvent labelling (as indicated).

Conclusion

Syndiotactic polystyrene/naphthalene forms fibrillar structures quite reminiscent of thermoreversible gels. From the phase diagram two compounds have been identified with different stoichiometries. The neutron diffraction study also supports the presence of polymer-solvent compound as derived from the phase diagram.

Acknowledgement

We gratefully acknowledge CEFIPRA(Grant No. 2808-2) for the financial support of the work. We are also very indebted to S. Zehnacker for carrying out the DSC experiments.

[1] Malanga, M. *Adv. Mat.* **2002**, *12*, 1869.
[2] Guerra, G.; Vitagliano, V.M.; De Rosa, C.; Petraccone, V.; Corradini, P. *Macromolecules* **1990**, *23*, 1539.
[3] Daniel, C.; De Luca, M.D.; Brulet, A.; Menelle, A.; Guenet, J.M. *Polymer* **1996**, *37*, 1273.
[4] Daniel, C.; Brulet, A.; Menelle, A.; Guenet, J.M. *Polymer* **1997**, *38*, 4193.
[5] Rudder, J.D.; Berghmans, H.; Schryver, F.C.D.; Basco, M.; Paoletti, S. *Macromolecules* **2002**, *35*, 9529.
[6] Ray, B. ; Elhasri, S. ; Thierry, A. ; Marie, P.; Guenet, J.M. *Macromolecules* **2002**, *35*, 9730
[7] Guenet, J.-M. *Thermoreversible geltaion of polymers and biopolymers*; Academic press: London, **1992**.
[8] Grassi, A . ; Pellechia, C. ; Longo, P. ; Ammendola, P. ; Zambelli, A. *Gazz. Chim. Ital.* **1987**, *19*, 2465.
[9] Point, J.J. ; Damman, P. ; Guenet, J.-M. *Polym. Commun.* **1991**,*32*, 477.
[10] Klein, M.; Menelle, A.; Mathis, A.; Guenet, J.-M. *Macromolecules* **1990**, *23*, 4591.

Macromol. Symp. **2005**, *222*, 81-86

Multiscale Porosity from Thermoreversible Poly(vinylidene fluoride) Gels in Diethyl Azelate

Debarshi Dasgupta, Arun K. Nandi*

Polymer Science Unit, Indian Association for the Cultivation of Science, Jadavpur, Kolkata-700032, India
E-mail: psuakn@mahendra.iacs.res.in

Summary: Poly(vinylidene fluoride) (PVF_2) produces thermoreversible gels in a series of diesters. The polymer-solvent complexation occurred for intermittent number of carbon atoms $n \geq 2$ and the enthalpy of complexation increased with increasing n. The gels were dried by replacing the diesters with low boiling solvent like cyclohexane (bp. 80 ^{0}C) and methylcyclohexane (bp. 99 ^{0}C). The porosity of the dried gels was measured using Poremaster-60. For PVF_2-DEAZ gel meso and macro porosity have been observed. The former pore dimensions have been attributed for polymer-solvent complexation while the macroporosity has been attributed for caging of solvent between the PVF_2 fibrils The porosity measured from nitrogen adsorption isotherms using BJH method indicate presence of minimum pore diameter of 3.8 nm for the 10% dried gel of PVF_2.

Keywords: BJH isotherm; macroporosity; mercury intrusion porosimetry; mesoporosity; thermoreversible gel

Introduction

Porous materials are vibrant areas of research because of their potential applications in sorption, catalysis, dielectric materials and separation process [1]. Porosity of materials is usually classified as microporous, mesoporous and macroporous according to their pore sizes. For example, microporous materials have pore diameter 0.5 nm–2 nm, mesoporous have pore diameter 2 nm–50 nm and macroporous substances have pore diameter > 50 nm. So far porosity development in inorganic materials is well studied but porosity development in organic materials is not well documented [1]. Some workers are using supramolecular bonding to prepare organic porous materials [2,3], however in such cases the porosity is limited to a particular value of micro or mesoporosity. But it may be needed particularly for the separation processes, a porous material with multiscale porosity, so that ions of different sizes can be seperated from industrial effluents.

To solve this problem we use thermoreversible polymer gels as in such gels solvents are trapped in different dimensions in polymer-solvent complexes and in the cages of the fibrillar network. The former may yield micro and meso porosity while the latter yields macroporosity.

DOI: 10.1002/masy.200550408

To achieve this goal we have used an important polymer poly(vinylidene fluoride) (PVF$_2$) which is highly used as membrane (milipore). PVF$_2$ produces thermoreversible gels in diesters, the gel morphology, however, depends on the number of intermittent carbon atoms of the diesters [4,5]. PVF$_2$-diester gels with n ≥ 2 produce fibrillar network morphology and also produce polymer-solvent complexes. The enthalpy of complexation increases with increase in 'n' and PVF$_2$-DEAZ gel exhibit highest enthalpy of complexation in the series studied so far [5]. We have tried to prepare multiporous polymeric materials of the diesters as the polymer-solvent complexes have better stability in these gels the results are included here for the PVF$_2$-DEAZ system.

Experimental

Poly(vinylidene fluoride) (PVF$_2$) is a product of Aldrich Chemical Co. USA. The weight average molecular weight ($\overline{M}_w$) of the sample is 1,80,000 g/mol and polydispersity index is 2.54 as obtained from GPC. The PVF$_2$ sample was recrystallized from its 0.2% solution in acetophenone, washed with methanol and was finally dried in vacuum at 60 ^{0}C for three days. The diester diethyl azelate was purchased from Lancaster, England and was used as received.

The gels were prepared by taking appropriate amount of polymer and solvent in glass tubes (8 mm in diameter) and were degassed by repeated freeze-thaw technique. They were then sealed in vacuum (10^{-3} mm Hg) and were made homogeneous at 180 ^{0}C with occasional shaking. The tubes were then quenched to 25 ^{0}C (room temperature) to produce the gel. To produce the porous material from the gel the gel was taken out in a petry dish and it was fully dipped with cyclohexane. The solvent was replaced by fresh batches to drive the replacement equilibrium faster in every 12 hours. Such process was repeated for 3-4 days and was tasted for complete replacement of ester by pouring in a white paper. Absences of oily spot indicate absence of diester in the PVF$_2$ matrix. The cyclohexane was poured off and the film was dipped into methanol for two days with three to four times change by methanol. This methanol replacement in necessary as without methanol a small amount of >C=O peak intensity is observed in FT IR spectra of cyclohexane dried PVF$_2$ gel. Finally the films were dried in vacuum at 60 ^{0}C for three days. Complete drying from diesters were confirmed by the absence of carbonyl peak (1740 cm^{-1}) in the FTIR spectra of the dried gels.

Porosity of the samples was measured both by mercury intrusion porosimetry and also by N$_2$ adsorption isotherm using Barrett-Joyner-Halenda (BJH) method. The mercury intrusion porosimetry has been done using the instrument Poremaster 60, Quantachrome instruments

USA and BJH isotherms were done using SA 3100 Surface Area and Pore Size Analyzer instrument (Beckman Coulter).

The melting point and enthalpy of fusion of the dried gels were measured using a differential scanning calorimeter DSC-7 (Perkin Elmer) under N_2 atmosphere. The samples were heated from 50 ^{0}C at the rate of 40 ^{0}C/ min to 227 ^{0}C. The higher heating rate was chosen to avoid any melt recrystallization of the PVF_2 sample.

Result and Discussion

In Fig. 1(a) and Fig. 1(b) the high pressure and low pressure mercury intrusion histograms of the dried PVF_2-DEAZ gels are presented.

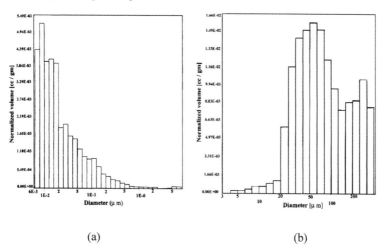

(a)   (b)

Figure 1. Mercury intrusion histograms of dried PVF_2 – DEAZ gel (10% w/v) (a) high pressure (34000 PSI) and (b) low pressure (50 PSI).

It is apparent from the figure 1(a) that substantial amount of pore volume exist of pore diameter 6-50 nm. It also indicates that there may remain pore of smaller diameter than 6 nm. figure 1(b) indicates there is substantial amount of macropore (diameter 20μm – 200 μm) in the sample. Thus from figure 1 it is now established that dried PVF_2-DEAZ gel have both meso porosity macro porosity. In figure 2 the N_2 adsorption isotherm and pore size distribution (BJH) of dried PVF_2-DEAZ gel is presented.

It is clear from the figure that pore volume has a maxima at pore diameter 6.14 nm. It indicates the majority of mesopores have diameter of ~ 6 nm. The minimum pore diameter observed from BJH method is 3.8nm. Thus from figure 1 and 2 we may conclude that the dried PVF$_2$-DEAZ gel have both meso and macroporosity. In a word this gel has multiporous structure. In figure 3 the hysteresis loop of intrusion and extrusion volume of mercury is presented.

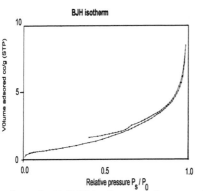
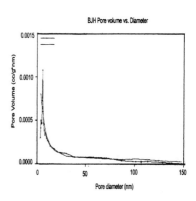

Figure 2. (a) BJH isotherm of nitrogen at 77° K and (b) pore size distribution in mesoporous PVF$_2$ film obtained after drying 10%(w/v) PVF$_2$ - DEAZ gel.

It is clear from the figure that pore volume has a maxima at pore diameter 6.14 nm. It indicates the majority of mesopores have diameter of ~ 6 nm. The minimum pore diameter observed from BJH method is 3.8nm. Thus from figure 1 and 2 we may conclude that the dried PVF$_2$-DEAZ gel have both meso and macroporosity. In a word this gel has multiporous structure. In figure 3 the hysteresis loop of intrusion and extrusion volume of mercury is presented

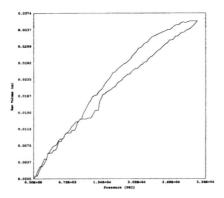

Figure 3. Hysteresis loop of 10% dried PVF$_2$ – DEAZ gel during mercury intrusion (lower) and extrusion (upper) processes at 77°K.

It indicates higher volume of Hg during extrusion at the same pressure. These results signify that there is some channel structure between the pores [6]. These channels preserve some mercury yielding higher volume during extrusion under same pressure.

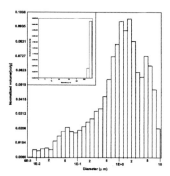

Figure 4. High pressure mercury intrusion histogram of 25% PVF$_2$ –DEAZ gel (Inset low pressure histogram of same sample).

In figure 4 the high-pressure histogram of 25% PVF$_2$-DEAZ gel is presented. It is apparent from the figure that there is both mesoporosity and macroporosity but the relative population of mesoporosity is much lesser than that of 10% gel. The low pressure histogram is shown in the inset of figure 4 where a small fraction of macropores (diameter >200 µm) is detected. It means the macropores are of lower diameter than those obtained from 10% gel. In a word there is a continuous distribution of pores of diameter 6 nm to 10 µm. The histeresis diagram is very much similar to that of 10% PVF$_2$-DEAZ gel indicating interconnectivity between the pores.

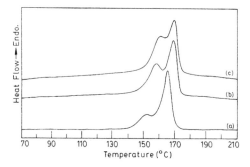

Figure 5. DSC melting endotherms of the dried PVF$_2$ - DEAZ gels at the heating rate of 40^0/min (a) 0%, (b) 15% and (c) 25% (w/v).

In figure 5 the DSC thermogramms of dried PVF_2-DEAZ gel of three different compositions are presented. It is apparent that there are two melting peaks. They may be due to two different type of PVF_2 crystals present in the dried gels as there is no possiblity of melt recrystallization in such a higher heating rate. The reason for the two melting peaks is not yet clear. Both melting point and enthalpy values increase with increasing the polymer concentration. A probable explanation is that with increasing PVF_2 concentration the pore volume decreases and that increases the crystalline thickness causing the increase of melting point and enthalpy of fusion.

Conclusion

It may be concluded from these results that polymer materials of multiscale porosity can be prepared by drying the thermoreversible gel in appropriate manner. There is connectivity between the pores and with increasing polymer concentration the pore size distribution becomes narrower.

Acknowledgement

We gratefully acknowledge IFCPAR Grant No. 2808-2 for financial support of the work. We also acknowledge Prof. P. Bhargav and Prof. P. Paramanik of IIT Kharagpur for their help in porosity measurement.

[1] C. N. R. Rao *Bull. Mater. Sci.* **1999**, *22*, 141.
[2] V. R. Pedireddi; S. Chatterjee; A. Ranganathan; C. N. R. Rao *J. Am. Chem. Soc.* **1997**, *119*, 10867.
[3] A. Ranganathan; V. R. Pedireddi; C. N. R. Rao *J. Am. Chem. Soc.* **1999**, *121*, 1752.
[4] A. K. Dikshit; A. K. Nandi *Macromolecules* **1998**, *31*, 8886.
[5] A. K. Dikshit; A. K. Nandi *Macromolecules* **2000**, *33*, 2616.
[6] C. A. Leon; Y. Leon *Adv. Colloid Interface Sci.* **1998**, *76-77*, 341.

Macromol. Symp. **2005**, *222*, 87-92

Photo-Deformation of Syndiotactic Polystyrene Gels

Hideyuki Itagaki,* Reiko Iida, Jun Mochizuki

Department of Chemistry, Graduate School of Electronic Science and Technology, Shizuoka University, 836 Ohya, Shizuoka 422-8529, Japan
Fax: (+81) 54 2373354; E-mail: itagaki@ed.shizuoka.ac.jp

Summary: We report a usual and novel phenomenon that an excitation light beam of a spectrofluorometer produced a black mark in gels of syndiotactic polystyrene (SPS). This mark whose shape is identical with that of an excitation light beam was found to appear in SPS/chloroform gel just after only irradiation at 260 nm for 5 to 10 min. The heating of the gel with the mark at 80°C diminished the mark and recovered uniformly the gel to be as it had been before irradiation. We discuss what the black mark is and why it is produced by light irradiation.

Keywords: gels; light irradiation; networks; polymer-solvent compounds; sols

Introduction

Syndiotactic polystyrene (SPS) is known to form thermoreversible gels, and more than 60 papers on SPS gels have been published so far since Kobayashi et al[1]. SPS chains forming network in a gel state (Figure 1) are already established to have a structure of polymer-solvent compounds[2,3] and/or clathrate crystal structures.[4,5]

In order to monitor free volume among SPS chains in gel form, we applied the fluorescence probe method to SPS gel systems in the same method as we did for isotactic polystyrene gel systems[6,7]: we dispersed naphthalene (NP), 1-methylnaphthalene (MN), 1,5-dimethylnaphthalene (DMN), and anthracene (A) in the SPS gels and examined in detail their fluorescence anisotropy values while changing the concentration of SPS. During these studies, we found the novel phenomenon that an excitation light beam of a spectrofluorometer can make a black mark in gels of SPS (Figure 2), although the excitation light beam is weak. When the total time of the measurements was short, the mark looked like a black line. However, the shape of the black mark was found to be identical with that of an excitation light beam, when the measurement time was long.

The objective of the present work is to clarify (i) what the black mark observed is and (ii) why this black mark is produced by the irradiation of weak light source.

          DOI: 10.1002/masy.200550409

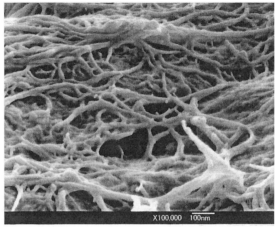

Figure 1. Scanning electron microscopy pictures of SPS/chloroform gel. The bar indicates 100 nm.

Experimental

The SPS sample used in the present study was kindly provided by Idemitsu Oil Co.: the weight average molecular weight is 1.52×10^5 (Mw/Mn=1.9). Chloroform, bromoform, and toluene were used as solvents. NP, MN, DMN, and A were used without further purification. All the gels used for the light irradiation were directly prepared in hermetically sealed quartz cells with optical path length of 1 mm and 1 cm by heating solvents and SPS until they became complete solutions and then cooling them at 4°C for 1 day. The gels were irradiated by either the light source of spectrofluorometer (resolved by slits: 5.9×10^{16} s^{-1} at 260 nm) or 500 W xenon light (resolved by interference filters: 1.6×10^{18} s^{-1} at 263 nm): the infrared light of the xenon lamp was removed by using mirrors.

Figure 2. SPS/chloroform gel prepared in a quartz cell after the fluorescence measurements.

Figure 3. The front and side views of the 0.87%(wt/wt) SPS/ chloroform gel prepared in a 1 cm × 1 cm quartz cell after being irradiated at 263 nm for 4 hours and left at -25°C overnight.

Results and discussion

In order to clarify what the black mark produced in SPS gels after the fluorescence measurements is (Figure 2), we irradiated an SPS/chloroform gel for a longer time with a stronger light source. Figure 3 shows the SPS/chloroform gel that was irradiated for 4 hrs at 263 nm (1.6×10^{18} quanta/sec) using a 500 W xenon lamp and kept at -25°C overnight. It clearly demonstrates that a hole was formed in the SPS/ chloroform gel by light irradiation. Bubbles sometimes came out during keeping the gels in a freezer, but did not always appear. Consequently, the mark in an SPS gel produced by the irradiation turned out to be a hole, namely of SPS solution form. Since the infrared beams of the xenon lamp were cut thoroughly, this phenomenon was not induced by thermal heating.

The heating of the gel at 80°C diminished the hole, and the perfect gel was reformed when cooled down. We do not claim strongly that this would be efficient as a memory, but we can repeatedly write something onto SPS/chloroform gels by using uv light and reset it by heating them.

Next we examined the best condition to produce holes in SPS gels in order to clarify the process to give a sol form only by irradiation of weak light. We changed the conditions to prepare SPS gels such as concentration of SPS, additives of fluorescent probe molecules, solvents of gelation, and also changed the conditions to irradiate gels such as excitation

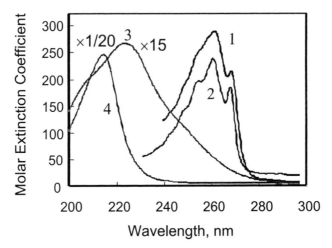

Figure 4. Uv absorption spectra of SPS/chloroform gel (1), toluene in cyclohexane (CH) (2), bromoform in CH (3), and chloroform in CH (4).

wavelength, intensities of excitation beam and irradiation time.

Because this unusual phenomenon was observed first for SPS/chloroform gels containing fluorescent probe molecules such as NP and MN, the influence of the addiives was examined in detail. However, in the long run, holes were formed more quickly and efficiently in the SPS/chloroform gels without any additives. Concerning the solvents of SPS gels, chloroform, bromoform, and toluene were examined whether they can form gels giving birth to holes when irradiated. Among three solvents, only chloroform gives a gel forming holes by uv light irradiation. Moreover, in the case of SPS/chloroform gels prepared in a quartz cell with optical path length of 1 mm, holes appeared at SPS concentrations being lower than 4 %(wt/wt): namely the concentration of SPS should not be too thick in order to produce holes when irradiated.

These experimental results clearly suggest that the formation of holes strongly depends on how much phenyl groups of SPS can absorb photons without being interfered with other molecules. Figure 4 shows the uv spectrum of SPS/chloroform gel together with the spectra of three compounds used as solvents. The solvents whose absorption is overlapping with the spectrum of SPS are found not to give SPS gels producing a hole when irradiated. Only chloroform has almost no absorbance in the wavelength range where phenyl groups of SPS have higher absorption. When the concentration of SPS is high, the light is considered unable to go into the inside of the gels, and the hole is not produced.

In order to make sure that light absorption of phenyl groups of SPS is most important for forming a hole at irradiation, we examined the dependence of excitation wavelength on formation of holes in 0.4 %(wt/wt) SPS/chloroform gels prepared in quartz cells with optical path length of 1 mm by using xenon lamp resolved by interference filters. Finally the hole was formed most quickly and sharply when the gel was excited at 263 nm, which is the peak wavelength of SPS absorption. When the gel was excited at 260 and 270 nm, the formation of a hole was also efficient. However, when the excitation wavelengths are 280 and 290 nm, it took quite a long time to form a hole, and at wavelengths higher than 300 nm, no holes were observed even after irradiation for 8 hours. In summary, the efficiency of forming a hole in an SPS gel corresponds to the absorption spectrum of SPS (Figure 4). Finally, the present novel process was concluded to start with the uv light absorption by side-chain phenyl groups of SPS.

How does this process take place after the excitation of phenyl groups of SPS gels? When a side-chain phenyl group is excited, three processes are possible: (i) the excited phenyl group deactivates by emitting fluorescence, (ii) it deactivates nonradiatively by vibrational motion, and (iii) it is photodegraded into a non-aromatic compound.

In the case of (i), the excitation energy released from SPS cannot be absorbed by chloroform, a solvent molecule, since the absorption spectrum of chloroform does not overlap with the fluorescence spectrum of SPS. Accordingly, nothing is expected to happen. However, the process of (ii) and (iii) should affect a polymer-solvent compound of SPS gel. The excitation energy of a side-chain phenyl group taking part in a polymer-solvent compound would be transferred to a chloroform molecule in the molecular compound by way of vibrational motion. This energy would urge the motion of the chloroform and may dissociate the polymer-solvent compound partly. If an excited phenyl group is degraded (process (iii)), the adjacent chloroform in the same polymer-solvent compound would become mobile because the degraded phenyl group could never keep the chloroform molecule fixed in this molecular compound any more.

Figure 5 shows the initial change of the uv spectra of 0.40% SPS/chloroform gel when it was irradiated at 263 nm. It is clear that photodegradation occurs a bit, but the decrease of phenyl groups due to the degradation is not so much. Nevertheless, the formation of a hole took place within 5 min. Thus, the process (ii) is more probable.

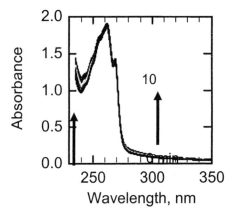

Figure 5. Change of uv spectra of 0.40% SPS/chloroform gel with an increase of irradiation time (every each 1 min from 0 to 10 min). The cross section part of the gel was perfectly the same for both irradiation at 263 nm and measurements of uv spectra.

So far no other similar phenomena induced by light irradiation have been published, but Ramzi et al. reported that agarose aggregates in water/DMSO underwent disruption while performing electric birefringence experiments[8]. It is probable that the electric field or the heat perturbs the molecular organization between solvent and polymer within the physical junctions by disorienting solvent molecules, resulting in destabilizing the organization.

Conclusion

A novel phenomenon was discovered that a hole is produced in SPS/chloroform gels by uv light irradiation. The heat released by nonradiative deactivation of excited side-chain phenyl groups of SPS is assumed to accelerate the motion of the adjacent solvent molecule in the polymer-solvent compound. This motion is concluded to urge the dissociation of the polymer-solvent compound. It would loosen and untie several networks and make the part of gels irradiated to be in sol form , followed by the destruction of the gel structure.

[1] M. Kobayashi, T. Nakaoki, N. Ishihara, *Macromolecules* **1990**, 23, 78.
[2] Ch. Daniel, M. D. Deluca, J.-M. Guenet, A. Brulet, A. Menelle, *Polymer* **1996**, 37, 1273.
[3] Ch. Daniel, A. Menelle, A. Brulet, J.-M. Guenet, *Polymer* **1997**, 38, 4193.
[4] Ch. Daniel, G. Guerra, P. Musto, *Macromolecules* **2002**, 35, 2243.
[5] C. S. J. van Hooy-Corstjens, P. C. M. M. Magusin, S. Rastogi, P. J. Lemstra, *Macromolecules* **2002**, 35, 6630.
[6] H. Itagaki, Y. Nakatani, *Macromolecules* **1997**, 30, 7793.
[7] H. Itagaki, *Macromol. Symp.* **2001**, 166, 13.
[8] M. Ramzi, E. Mendes, C. Rochas, J.-M. Guenet, *Polymer* **2000**, 41, 559.

Methods to Analyse the Texture of Alginate Aerogel Microspheres

Romain Valentin, Karine Molvinger, Françoise Quignard, Francesco Di Renzo*

Laboratoire Matériaux Catalytiques et Catalyse en Chimie Organique, UMR 5618 ENSCM-CNRS-UM1, ENSCM, 8 Rue de l'Ecole Normale, 34296 Montpellier, France
Email: direnzo@cit.enscm.fr

Summary: Nitrogen adsorption at 77 K has been applied to the study of the texture of alginate aerogel microspheres obtained by CO_2 supercritical drying of alcogels. The limited volume shrinkage suggests that the aerogels preserve the texture of the hydrogels. Alginate aerogels presents a N_2 adsorption at small pressure higher than reference non-porous silica, to be attributed to the polarity of the surface or to a small microporous volume. The aggregated nanobead strings of the guluronic-rich gels accounts for a significant mesoporosity. The N_2 adsorption results correspond to electron microscopy observations for features smaller than 50 nm.

Keywords: aerogel; alginate; biopolymer; nitrogen adsorption; supercritical drying; textural characterisation

Introduction

Alginates are abundant polysaccharides produced by brown algae, mainly composed of (1-4) linked β-D-mannuronic (M) and α-L-guluronic (G) residues (figure 1), in varying proportions, sequence and molecular weight. The use of alginate in most applications lies in its ability to form heat-stable strong gels with divalent or trivalent cations, most generally Ca^{2+}. The strength of the gel has been attributed to the electrostatical interaction of the cation with the guluronic residues, affording a three-dimensional network described by an "egg-box" model.[1] Alginates are used as thickeners in the food industry as well as for the encapsulation of bioactive materials like drugs,[2] proteins,[3] living cells,[4] and enzymes.[5]

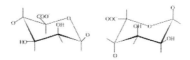

Figure 1. The alginate monomers: (left-hand) β-D-mannuronate (**M**) and (right-hand) α-L-guluronate (**G**).

DOI: 10.1002/masy.200550410

The structure of alginate hydrogels has been widely studied by scanning electron microscopy. Structural features in the size range 1-100 μm have been observed[6, 7] and early attributed to modifications of the gel due to the drying process.[8] The presence of large cavities in the dried material has been confirmed by neutron diffusivity measurements[9] but it is still subject of debate at which point structural features like micrometric shafts or crevices are already present in the hydrogel.[9, 10] The presence of a smaller porosity with a size scale of tens of nanometers is evidenced by the easy accessibility of the alginate hydrogels, through which macromolecules up to 20,000 Da can freely diffuse.[11, 12] This porosity seems to be delimited by a network of fibrillar structural elements with a diameter in the range 20-50 nm, as evidenced by scanning electron microscopy.[13-15]

Pore size of materials with cavities smaller than 50 nm and surface area can be easily measured by nitrogen adsorption methods.[16, 17] In this work, such techniques, typically used for the characterization of adsorbents or inorganic catalysts, have been applied to the study of the texture of alginate aerogels. Supercritical CO_2 drying of alcogels has been suggested as the best method to obtain an image of the wet materials in the solid state. This procedure releases the porous texture quite intact by avoiding the pore collapse phenomenon.[9, 18]

An investigation of the textural properties of alginate aerogels of three different compositions in term of mannuronic/guluronic ratio is reported. Scanning electron microscopy observations have been associated to the analysis of N_2 sorption isotherms.

Experimental

The characteristics of the three different sodium alginates used in this work are reported in table 1.

Table 1. Characteristics of the alginates samples.

Sample	Guluronic percent (by ^{13}C NMR)	viscosity 2 % solution (w/v) $\eta(mPa.s^{-1})$	average Mw
G-20	20	200	200 000
G-45	45	2000	400 000
G-76	76	2000	400 000

Sodium alginate was dissolved in distilled water at a concentration of 2% (w/v). The polymer solution was added dropwise at room temperature to the stirred $CaCl_2$ (Aldrich) solution (0.24 M) using a syringe with a 0.8mm diameter needle. The microspheres were cured in the gelation solution for a given time. The maturation time, 1 hour for G-20, 3 hours for G-45 and G-76, was chosen as the time providing the highest surface area at the end of the preparation. The microspheres were then dehydrated by immersion in a series of successive ethanol-water baths of increasing alcohol concentration (10, 30, 50, 70, 90, 100 %) during 15 min each.[18] Then, the microspheres were dried under supercritical CO_2 conditions (74 bars, 31.5°C) in a Polaron 3100 apparatus.

Scanning electron micrographs (SEM) of cross-sections of the dried microspheres were obtained on an Hitachi apparatus after platinum metallization.

Nitrogen adsorption/desorption isotherms were recorded in a Micromeritics ASAP 2010 apparatus at 77 K after outgassing the sample at 353 K under vacuum until a stable $3 \cdot 10^{-5}$ Torr pressure was obtained without pumping. A reference isotherm was measured on fumed silica Aerosil 200 and used for comparison plots in the relative pressure range 0.05-0.93.

Results and Discussion

The properties of alginate gels are influenced by the ratio and sequencing of the uronic monomers,[19] the concentration of the cation in the maturation bath and the time of maturation.[20, 21] Three different alginates were used for this study, with different guluronic fractions: G-20 (20 % guluronic), G-45 (45 % guluronic) and G-76 (76 % guluronic).

Gel spheres obtained from alginate G-45 are represented in Figure 2 at different stages of preparation: hydrogel spheres (Figure 2a), alcogel spheres obtained after exchange of water by ethanol (Figure 2b), and aerogel spheres obtained after supercritical CO_2 drying (Figure 2c). For sake of comparison, the xerogel spheres obtained from the alcogel by ethanol evaporation at room temperature are also represented (Figure 2d). For all alginate samples, the average sphere diameter at the different stages of preparation is reported in Table 2.

Table 2. Average bead size and mean square error (mm) for three different alginates at different stages of preparation.

	G-20	G-45	G-76
hydrogel	3.03 ± 0.11	2.72 ± 0.17	2.76 ± 0.13
alcogel	2.76 ± 0.13	2.53 ± 0.13	2.63 ± 0.19
aerogel	2.81 ± 0.24	2.50 ± 0.14	2.35 ± 0.11
xerogel	0.87 ± 0.08	0.83 ± 0.03	0.90 ± 0.04

For all alginate samples, the exchange of ethanol for water brought about a size shrinkage lower than 10 %, corresponding to a volume shrinkage lower than 25 %, in good agreement with literature results.[18, 22] The supercritical CO_2 drying allows to form the aerogel from the alcogel and brings about no size change of the spheres of G-20 and G-45. The shrinking observed in the case of G-76 is at the limit of the mean square error and probably witnesses the experimental problems that sometimes affect supercritical drying.[21]

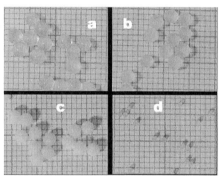

Figure 2. G-45 alginate beads at different phases of the preparation. (a) hydrogel, (b) alcogel, (c) aerogel from supercritical CO_2-dried alcogel, (d) xerogel from room temperature-dried alcogel. Grid square side 1 mm.

By contrast, the ethanol evaporation in room conditions forms a xerogel and brings about a dramatic shrinking of the alginate spheres. The size of the xerogel spheres is about 30 % of the size of the hydrogel spheres, hence the xerogel volume is 2.7 % of the volume of the hydrogel. This data has to be compared with the concentration of the initial alginate solution, 2 % (w/v), and indicates that the alcohol evaporation has produced a compact alginate with virtually no porous structure.

Scanning electron micrographs of cross-sections of the aerogels obtained from the three alginate samples are reported in Figure 3.

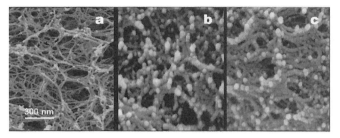

Figure 3. Scanning electron microscopy of cross-sections of alginate aerogels. (a) G-20, (b) G-45, and (c) G-76. Same scale for all pictures.

The texture of the G-20 aerogel is an open network of fibres with diameter 10-15 nm, in good agreement with previous reports.[10, 13, 15] The nanostructure of the G-45 and G-76 aerogels is instead characterized by important instances of aggregations and is constituted by strings of nanobeads with diameter 20-30 nm. Bead strings are frequently associated to form pillars and, in the case of G-76, sheets. The aerogels from guluronic-rich alginates present a more compact nanostructure than the mannuronic-rich sample, seemingly in agreement with the better mechanical properties of the guluronic-rich gels.[19]

Nitrogen adsorption-desorption isotherms provide useful information about several textural properties.[16, 17] The adsorption at low relative pressure allows to evaluate the surface area of the sample or the presence of microporosity. Microporosity can be differentiated from monolayer adsorption by drawing comparison plots, in which the volume adsorbed is plotted as a function of the volume adsorbed at the same pressure on a reference non-porous specimen. Several kinds of comparison plots are commonly used and the alpha-S plot, in which the unit value of the X axis corresponds to the volume adsorbed at relative pressure 0.40 on the reference isotherm, has been used in this communication. The presence of micropores is expected to be indicated by a positive extrapolated value of the comparison plot at alpha-S = 0.

Linear portions of the comparison plot correspond to pressure domains in which the adsorption takes place with the same mechanism as on the reference adsorbent, expectedly a monolayer-multilayer mechanism. Positive deviations from linearity corresponds to condensation phenomena and allows to quantitatively evaluate the presence of mesopores.

The surface area can be evaluated by the classical BET method or by the slope of the alpha-S plot. This last method is expected not to be affected by the presence of micropores.

The adsorption-desorption isotherms of N_2 at 77 K on the three alginate aerogels are presented in Figure 4 and the corresponding alpha-S plots are presented in Figure 5.

All isotherms of Figure 4 are type IV, typical of mesoporous solids with strong adsorbent-adsorbate interaction. The well-defined step at low relative pressure corresponds to a high energy of monolayer adsorption. This effect is confirmed by the high C_{BET} parameter, about 140, reported in Table 3. Such a strong low-pressure adsorption can easily hid the presence of micropores, and indeed the comparison plots extrapolated at alpha-S = 0 indicate the presence of some microporosity in all samples (Figure5).

The micropore volume, reported in table 3, is very small, the observed value of 0.014 cm^3 g^{-1} corresponding to the adsorption of a nitrogen molecule per 12 uronic units, and indicates that the alginate at 77 K is quite impervious to the N_2 molecules. If it is taken into account that the reference used in the comparison plot, a fumed silica, has not the same composition as the polysaccharide samples, the measured micropore volume could also be attributed to a different shape of the adsorption isotherm on the highly polar surface of the alginate.

Table 3. Textural data from N_2 adsorption isotherms at 77 K.

	S-20	S-45	S-76
BET surface area ($m^2\,g^{-1}$)	342	339	507
alpha-S surface area ($m^2\,g^{-1}$)	314	302	470
micropore volume ($cm^3\,g^{-1}$)	0.009	0.014	0.013
C_{BET}	138	144	142
calculated fibre diameter (nm)	12.7	13.3	8.5
calculated nanobead diameter (nm)	19.0	20.0	12.8

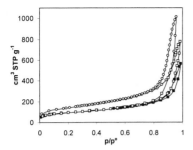

Figure 4. N_2 adsorption-desorption isotherms at 77 K for three alginate aerogels: (filled squares) G-20, (empty squares) G-45, (empty circles) G-76.

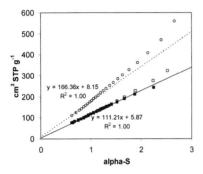

Figure 5. Alpha-S plots for three alginate aerogels: (filled squares) G-20, (empty squares) G-45, (empty circles) G-76. The linear parts of the plots for G-20 (full line) and G-76 (dotted line) are evidenced. The linear part of the G-45 curve, very similar to G-20, is omitted for sake of clarity.

The slow steady rise beyond $p/p°$ 0.1 of the isotherms of Figure 4 indicates the absence of any small mesoporosity. No hysteresis can be observed below relative pressure 0.8 which, in the case of cylindrical pores, would correspond to a smallest mesopore size limit of 13 nm.[23] At higher pressure, the mesopore distribution is open, with no solution of continuity with the macroporous domain.

In very open networks, as those presented by the samples under examination, only a limited fraction of the surface corresponds to the inner surface of mesopores and a reliable evaluation of the mesopore volume has not to take into account most of the adsorbate related to a monolayer-multilayer adsorption mechanism. In this case, the amount of condensate by capillarity can be easily measured on the comparison plots (Fig. 5) as the difference between the adsorbed volume and the extrapolation at high pressure of the linear portion of the alpha-S plot. In the case of the sample with the largest mesoporosity, the gel G-76, the volume adsorbed in mesopores with diameter smaller than 25 nm does not exceed 0.15 $cm^3 g^{-1}$.

In the case of gels G-20, the alpha-S plot reported in Figure 5 presents a slight but significant negative deviation at high relative pressure. This indicates that the sample presents less mesoporosity than the reference sample which, contrarily to expectations, presents some intergranular porosity and traces of hysteresis beyond p/p° 0.8.

The observed mesoporosity can be related to the gel morphology observed by electron microscopy. The gel G-20, virtually not mesoporous, is formed by an open network of isolated fibres. The gels G-45 and G-76, with significant mesoporosity, are formed by aggregated strings of nanobeads. It seems clear that the mesoporosity has to be attributed to cavities among aggregated strings, inside the strands which provide the mechanical stiffness of the guluronic-rich gels.

The BET surface area of the samples, reported in Table 3, is likely affected by the presence of microporosity. In this case, the slightly lower value calculated by the slope of the alpha-S plot is a more reliable evaluation of the surface area. This evaluation of the surface area, in the hypothesis of homogeneity of each sample, can provide some information on the size of the structural elements. It is easy to calculate the diameter of cylinders or spheres which would present the measured surface area, under the assumption of a density value for the material (unit density in this calculation). The calculated diameters are reported in Table 3. In the case of the sample G-20, a spherical model largely overestimates the size of the fibres, while a cylindrical model indicates a diameter 13 nm for the fibres, in good agreeement with electron microscopy results. In the case of the samples G-45 and G-76, a cylindrical model grossly underevaluates the size of the structural elements, while a spherical model approaches the size of the nanobeads observed by electron microscopy.

Conclusions

Nitrogen adsorption provides useful information for the characterisation of polysaccharide aerogels. Adsorption techniques alone provide quantitative data on surface area and microporous and mesoporous volume. The obtention of textural information from these data is only possible if independent information, provided by scanning electron microscopy, allows to choose between several possible models. The structure of alginate aerogels of different composition has been successfully determined by the use of these complementary techniques. The results suggest that CO_2 supercritical drying does not alter the texture of the original gel.

[1] G. T. Grant, E. R. Morris, D.A. Rees, P. J. Smith, D. Thom, *FEBS Lett.* **1973**, *32*, 195.

[2] M. Grassi, I. Colombo, R. Lapasin, *J. Control. Release* **2001**, *76*, 93.

[3] M. Kierstan, C. Bucke, *Biotechnol. Bioeng.* **2000**, *76*, 726.

[4] C. Stabler, K. Wilks, A. Sambanis, I. Constantinidis, *Biomaterials* **2001**, *22*, 1301.

[5] A. Blandini, M. Macias, D. Cantero, *Enzyme Microb. Technol.* **2000**, *27*, 319.

[6] S.C. Musgrave, N.W. Kerby, G.A. Codd, W.D.P. Stewart, *Eur. J. Appl. Microbiol. Biotechnol.* **1983**, *17*, 133.

[7] P. Scherer, M. Kluge, J. Klein, H. Sahm, *Biotechnol. Bioeng.* **1983**, *23*, 1057.

[8] D. Casson, A.N. Emery, *Enzyme Microb. Technol.* **1987**, *9*, 102.

[9] B.P. Hills, J. Godward, M. Debatty, L. Barras, C.P. Saturio, C. Ouwerx, *Magn. Res. Chem.* **2000**, *38*, 719.

[10] D. Serp, M. Mueller, U. Von Stocker, I.W. Marison, *Biotech. Bioeng.* **2002**, *79*, 253.

[11] H. Tanaka, M. Matsurama, I. Veliky, *Biotech. Bioeng.* **1994**, *26*, 53.

[12] M. Longo, I. Novella, L. Garcia, M. Diaz, *Enzyme Microb. Technol.* **1991**, *14*, 586.

[13] Y.M. Belatseva, V.B. Tolstognzov, D.B. Izymov, M.M. Genia, *Biofizika* **1972**, *17*, 744.

[14] F. Yokoyama, C. Achife, K. Takakira, Y. Yamashita, K. Monebe, *J. Macromol. Sci. Phys.* **1992**, *B31*, 463.

[15] D. Serp, M. Mueller, U. Von Stocker, I.W. Marison, *Biotech. Bioeng.* **2002**, *79*, 243.

[16] S. J. Gregg and K. S. W. Sing., *"Adsorption, Surface Area and Porosity"*, Academic Press, London 1982.

[17] F. Rouquerol, J. Rouquerol, K. Sing, *"Adsorption by powders and porous solids"*, Academic Press, San Diego 1999.

[18] A. Martinsen, I. Storrø, G. Skjåk-Bræk, *Biotech. Bioeng.* **1992**, *39*, 186.

[19] O. Smidsrød, *Faraday Discuss. Chem. Soc.* **1974**, *57*, 263.

[20] N. Velings, M.M. Mestdagh, *Polym. Gels Netw.* **1995**, *3*, 311.

[21] C. Ouverx, N. Velings, M.M. Mestdagh, M.A.V. Axelos, *Polym. Gels Netw.* **1998**, *6*, 393.

[22] B. Thu, O. Gåserød, D. Paus, A. Mikkelsen, G. Skjåk-Bræk, R. Toffanin, F. Vittur, R. Rizzo, *Biopolymers* **2000**, *53*, 60.

[23] J.C.P. Broekhoff, J.H. de Boer, *J. Catal.* **1968**, *10*, 377.

Macromol. Symp. **2005**, *222*, 103-108

Polyacrylamide-Grafted Silica as Special Type of Polymer-Colloid Complex

Olga Demchenko,* Tatyana Zheltonozhskaya, Svetlana Filipchenko, Vladimir Syromyatnikov

Macromolecular Chemistry Department, Kiev Taras Shevchenko National University, 64 Vladimirskaya Str., 01033, Kiev, Ukraine

Summary: Structure and properties of polyacrylamide-grafted silica have been investigated with the help of thermal analysis methods, NMR spectroscopy and by measuring the ability solubilization in benzene when compared with homopolymer polyacrylamide (PAA). More homogeneous structure, low rigidity and density of packing have been revealed for the polymer shell. PAA form a dense polymer shell, which interact with silica surface through H-bonds. ^{1}H NMR spectroscopy suggests no influence of silica particles on PAA stereoregularity. PAA-grafted silica considered as a special type of polymer-colloid complex where polymer chains are covalently bound to silica with one end and polymer segments along the chain are hydrogen bound to the particle surface.

Keywords: polyacrylamide-grafted silica; polymer shell; properties; structure

Introduction

Modification of inorganic surface through the use of physically adsorbed or grafted macromolecules constitutes an important means by which tailor both surface and polymer at interface properties governing, for instance, stabilization of colloidal dispersions adhesion, lubrication, biocompatibility.[1-4] A large body of knowledge is now available on polymers grafting onto nanoparticles.[5-8] Polymer-grafted particles synthesis usually includes the preliminary chemical modification of particle surface or polymer.[5,6,8] We have performed the radical graft polymerization of acrylamide (AA) onto silica nanoparticles using ceric ions. The unification of sorption ability of colloid particles surface and binding properties of long polymer chains in PAA-grafted silica produces its high flocculative capability during clarifying of model kaolin suspension (R=6,5 μm, C=30 kg m^{-3}) and at the coagulative-flocculative process of natural water clearing.[9,10] In the present work we consider some structure features and properties of PAA-grafted silica and peculiarities of PAA shell.

DOI: 10.1002/masy.200550411

Structure and properties of PAA-grafted silica

In a previous paper we have reported the grafting of PAA onto poly(vinyl alcohol) backbone, initiated with radicals formed by the redox reaction of ceric ion and alcoholic hydoxyl groups under nitrogen.[11] In the present paper to graft PAA onto silica the same cerium-ion-initiated solution polymerization technique have been used. AA polymerization in this process was initiated by redox system consisting of ceric ion and silanol group, which is naturally abundantly present on silica. Silica nanoparticles used was Aerosil A-175 with average hydrodynamic diameter of sol particles 22 nm. The molecular weight of grafts was estimated to be 1160000 from the intrinsic viscosity. The grafts number N was calculated to be on average 53 grafts per 1 silica particle from elementary analysis data, but considering the water content in polymer-grafted particle.[12] The confirmation of silica availability in grafted product was carried out by elemental analysis. Grafted product was investigated together with PAA sample ($\overline{M}_v$=1250000).

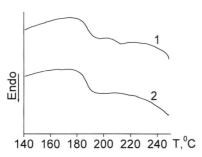

Figure 1. DSC thermograms for PAA (1) and PAA-grafted silica (2) (second run after heating to 160 °C and rapid cooling)

DSC thermograms obtained by differential scanning calorimetry (DSC) at the heating rate 16 °C/min with the help of Du Pont 1090 thermal analyzer are shown in Figure 1 and the data of thermal transitions are presented in Table 1. The glass transition temperature of PAA-grafted silica was determined to be lower than that for PAA but transition occurs in more narrow temperature interval (ΔT_g). This fact and the lack of ll-transition for PAA-grafted silica suggest more homogeneous structure, lesser rigidity and density of packing for PAA graft chains than for homopolymer.

Table 1. Thermal transition temperatures.

Name	T_g	ΔT_g	T_{ll} [1]	ΔT_{ll}	ΔC_p [2]
	°C	°C	°C	°C	$J \cdot g^{-1} \cdot K^{-1}$
PAA	188	13	206	14	0,65
PAA-grafted silica	186	11	-	-	0,67

[1] ll-transition temperature
[2] Heat capacity jump of α-relaxation transition

Thermal behavior of PAA-grafted silica was investigated by dynamic thermogravimetric analysis (DTA) in the non-isothermal regime on Q-1500 "quasi" technique of F.Paulik, G.Paulik, L.Erdey system at heating rate 2,5 °C/min. Weights of ~ 20 mg were used. Parameters of thermooxidative decomposition obtained from experimental TG, DTG and DTA curves are gathered in Table2. The common analogy between PAA-grafted silica and PAA decomposition in air that occurs in three stages is observed. But such differences for polymer-grafted particle exist: T_d (T_{in} of stage I) falls; ΔT of stage I, which accompanied by endothermic effect increases; ΔT of stage II, which accompanied by exothermic effect decreases; ΔT of stage III, which is accompanied by maximum exothermic effect extends; common temperature interval of decomposition falls. The thermostability reduction of filled polymers is well known.

Table 2. Parameters of thermooxidative decomposition.

Name	Stage	$T_{in} \div T_f$ [1]	ΔT [2]	T_{max} [3]	W [4]	Thermal effect	Residual
		°C	°C	°C	%	°C	%
PAA	I	256-344	88	302	16.5	endo	
	II	344-449	105	404	21.2	exo, 404	
	III	449-618	169	558	52.4	exo, 558	9.9
PAA-grafted silica	I	236-331	95	289	18.2	endo	
	II	331-406	75	376	18.1	exo, 374	
	III	406-587	181	535	60.3	endo	3.4

[1] Initial and final temperatures.
[2] Temperature interval.
[3] Temperature of the maximum decomposition rate.
[4] Weight loss.

Table 3. Formal kinetic parameters of two stages of thermooxidative decomposition.

Name	Stage	$T_{in} \div T_f$ [1]	E [2]	n [3]	Z [4]	k [5]
		°C	kJ/mol		1/s	1/s
PAA	I	256-344	143,9	1,15	$1,34 \cdot 10^{12}$	0,113
	II	344-449	165,2	1,40	$6,13 \cdot 10^{11}$	0,110
PAA-grafted silica	I	236-331	218,5	2,41	$5,96 \cdot 10^{19}$	0,292
	II	331-406	178,86	8,6	$6,32 \cdot 10^{61}$	42,77

[1] Initial and final temperatures.
[2] The activation energy.
[3] The order of reaction.
[4] The frequency factor in the Arrhenius equation.
[5] The rate constant of decomposition.

Formal kinetic parameters E, n, Z and k of two first decomposition stages determined by computer processing of experimental data from TG and DTG curves using Arrhenius equation in the matrix form and the least-squares method are presented in Table 3.[13]

Using literature and own experimental data we have explained the chemical transformations occurring during PAA decomposition in air previously.[14-16] The maximum values of formal kinetic parameters for PAA-grafted silica on the stage I of decomposition when comparing with PAA (Table 3) suggest interaction of grafts with silica through H-bonds, producing formation of separate cross-links between PAA graft chains and silanol groups at the temperature influence before the beginning of destruction. The largest values of formal kinetic parameters for PAA-grafted silica on the stage II of decomposition (Table 3) when the process of polymer oxidation occurs, suggest more weak interaction between PAA graft chains comparing with homopolymer producing high rate of the solubility and diffusion of oxygen in the polymer bulk.

Investigation of benzene solubilization by aqueous polymer-grafted particle solutions was

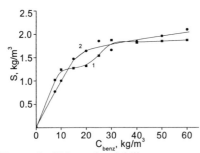

Figure 2. Value of solubilizaton *versus* benzene concentration with PAA (1) and PAA-grafted silica (2) aqueous solutions.

carried out by refractometry method.[17] The determination of values (Figure 2) and parameters (Table 4) of benzene solubilization was described previously.[17] Solubilization curve for PAA (Figure 2, curve 1) has two steps due to coiling of PAA macromolecules in the solubilization process at benzene concentration $C=22$ kg/m^3. Benzene binding with SiO$_2$ particles arises from formation of H-bonds between benzene π-electron system and

Table 4. Parameters of benzene solubilization with aqueous homopolymer and polymer-grafted particle solutions.

Name	S_{lim}	S_{red}	N [1]	V [2]
	kg/m^3	kg/m^3	mol$_{benz}$/mol$_{sample}$	nm^3
PAA	1.83	0.08	1505	135
SiO$_2$ [3]	1.89	0.14	4450[4]	-
PAA-grafted silica	1.70	-	-	-

[1]Number of benzene molecules bound by one sample macromolecule.
[2]Size of hydrophobic regions.
[3]Literature data. [18]
[4]Number of benzene molecules bound with one SiO$_2$ particle.

silanol groups of silica surface. The absence of surplus benzene solubility in PAA-grafted silica aqueous solution arises from destruction of hydrophobic regions in polymer shell due to PAA interaction with particles surface and formation of dense polymer shell, which prevents benzene transport to SiO_2 surface.

^{1}H NMR spectra were recorded with a Varian Mercury-400 spectrometer operating at 400 MHz. The elucidation of PAA homopolymer stereoregularity (^{13}C NMR spectroscopy) has been performed previously using literature data where authors obtained good resolved ^{13}C NMR spectrum for low-molecular-weight PAA.[19,20] The resulting spectrum showed the methylene, methine and carbonyl carbons of head-to-tail vinyl polymer. Polyacrylamide obeys Bernoulli statistics with $P_m = 0.43$, which is not unlike other vinyl polymers. As we have not obtained the good resolved ^{13}C NMR spectrum for homopolymer, PAA-grafted silica was investigated by ^{1}H NMR spectroscopy (Figure 3).[20] From the overall line shapes and peak percentage contribution to intensity of methine and methylene ^{1}H signals in the limits of integrating errors it is evident that the results obtained for PAA hold for PAA-*grafted*. Insignificant differences in peak positions are observed due to their measurements

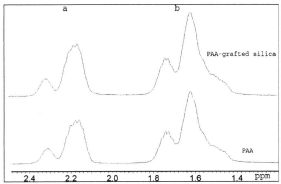

referencing to H_2O. So the microstructures of polymer formed at homopolymerization and at grafting polymerization onto silica are not differed. The possible explanation for this is few contacts of growing PAA chains with SiO_2 when only sufficiently long growing chain part contact with silica. It is possible when the interaction is week or when the steric hindrances exist. But it is well

Figure 3. Methine (a) and methylene (b) ^{1}H resonances of PAA (0.01 g/cm[3] solution in D_2O) and PAA-*grafted* in PAA-grafted silica (0.001 g/cm[3] solution in D_2O).

known about great affinity of PAA to silica surface. So due to the steric hindrances there are few contacts of growing PAA chain with SiO_2 and sufficiently long unbound parts of grafted macromolecule.

Conclusion

The main conclusions from our studies are: (i) more homogeneous structure, lesser rigidity and density of packing are observed for PAA-grafted than for PAA; (ii) the cooperative system of H-bonds exist between silica surface and PAA which form the dense polymer shell; (iii) the stereoregularity of PAA-grafted and homopolymer are not differed. Graft polymerization of PAA onto silica particles results in polymer-colloid complex formation of special type where polymer chains are covalently bound to silica with one end, where are a few contacts of grafted chains with particle surface through H-bonding and sufficiently long unbound parts of grafted chains (loops and tails) which interacts between themselves (Figure 4).

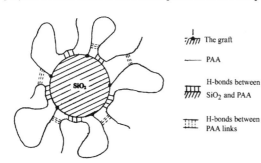

The graft

—— PAA

H-bonds between
SiO$_2$ and PAA

H-bonds between
PAA links

Figure 4. Schematic model for PAA-grafted silica.

[1] N.Tsubokawa, T.Umeno, *J. Jap. Soc. Colour. Mater.* **1991**, *64*, 751.
[2] A.Dorinson, K.C.Ludema, *"Mechanisms and Chemistry in Lubrication"*, Elsevier, Amsterdam 1985.
[3] T.Okano, T.Aoyagi, K.Kataoka, K.Abe, Y.Sakurai, M.Shimada, *J. Biomed. Mater. Res.*1986, *20*, 919.
[4] M.I.Baraton, F.Chancel, L.Merhari, *Nanostruct. Mater.* 1997, *9*, 319.
[5] M.L.Rong, Q.L.Ji, M.Q.Zhang, K.Friedrich, *Europ. Polym. J.* 2002, *38*, 1573.
[6] Y.Shirai, N.Tsubokawa, *Reactive and Functional Polym.* 1997, *32*, 53.
[7] J.Liu, R.Pelton, A.N.Hrymak, *J.Coll. Int. Sci.* 2000, *227*, 408.
[8] Y.Shirai, K.Kawasura, N.Tsubokawa, *Progr. In Organic Coat.* 1999, *36*, 217.
[9] V.V.Goncharuk, I.M.Solomentseva, V.F.Skubchenko, T.B.Zheltonozhskaya, O.V.Demchenko, N.V.Kutsevol, V.G.Syromyatnikov, V,S.Bilyk, V,M.Olenchenko, UA 44968A **2002**.
[10] V.V.Goncharuk, I.M.Solomentseva, V.F.Skubchenko, T.B.Zheltonozhskaya, O.V.Demchenko, N.V.Kutsevol, V.G.Syromyatnikov, UA 36528 **2003**.
[11] B.V.Eremenko, M.L.Malysheva, O.D.Rusina, N.V.Kutsevol, T.B.Zheltonozhskaya, *Colloids Surfaces A: Physicochem Eng. Aspects.* **1995**, *98*, 19.
[12] C. L.McCormik, L. S. J.Park, *Appl. Polym. Sci.* **1985**, *30*, 45.
[13] O.V.Suberlyak, G. Yu.Uigeliy, *Dopovidy NAN Ukr.* **1998**, *11*, 150.
[14] W. M.Leung, D. E. J.Axelson, *Polym. Science.* **1987**, *25*, 1825.
[15] N. V.Kutsevol, T. B.Zheltonozhskaya, O. V.Demchenko, L. R.Kunitskaya, V. G.Syromyatnikov, *Vysokomol. Soed.* **2004**, *A46(5)*, 839.
[16] O.Demchenko, T.Zheltonozhskaya, J.-M.Guenet, S.Filipchenko, V.Syromyatnikov, *Macromol. Symp.* **2003**, 203,183.
[17] N.Zagdanskaya, L.Momot, T.Zheltonozhskaya, J.-M.Guenet, V.Syromyatnikov, *Macromol. Symp.* **2003**, 203,193.
[18] T.B.Zheltonozhskaya, O.O.Romankevich, N.V.Kutsevol, V.G.Syromyatnikov, *Proceed. of the 2th International Conf. On Carpathian Euroregion Ecology,* **1997**, Miskolc (Hungary), 149.
[19] J. E.Lancaster, , M. N. J.O'Conner,*Polym. Science: Polymer Letters Edition.* **1982**, *20*, 550.
[20] O.Demchenko, T.Zheltonozhskaya, A.Turov, M.Tsapko, V.Syromyatnikov, *Mol. Cryst. and Liq. Cryst.,* *submited for publication.*

Gelation of Misfolded Proteins

Aline F. Miller

Department of Chemical Engineering, UMIST, P.O. Box 88, Manchester, M60 1QD, UK
Email: a.miller@umist.ac.uk

Summary: Insulin protein was exposed to mildly denaturing conditions (heat and low pH) to encourage the formation of beta-sheet rich amyloid fibrils. This resulted in an increase in viscosity of our protein samples and the morphology and thermodynamics of the resulting hydrogel were monitored using environmental scanning electron microscopy and micro differential scanning calorimetry respectively. It was found that the beta-sheet fibrils aggregated further to form macrofibrils, 2 μm in diameter and several microns in length. These long, flexible macrofibrils became entangled to form hydrogels with controllable mesh size: the higher the incubation temperature the higher the number of entanglements, and consequently the smaller the mesh size.

Keywords: amyloid fibril; fibrillar hydrogel; self-assembly

Introduction

Protein folding has become the focus of much attention world-wide because of its implication in a wide range of diseases including v-CJD and Alzeimers'.[1,2] The basic steps are being elucidated, with the recognition that probably all proteins are capable – under appropriate conditions – of mis-folding and subsequently self-assembling into the beta-sheet rich filaments that make up an amyloid fibril.[3] These fibrils consist of several protofilaments intertwined in a helical fashion to form a very stable, internally hydrogen -bonded fibril, which is typically nanometres in width but microns in length.[4] In laboratory experiments, it is common to encourage proteins to form fibrils by exposing the (correctly) folded protein to mildly denaturing conditions, such as low pH or elevated temperatures. As denaturation and self-assembly proceeds, the interaction of individual amyloid fibrils, which is considered to be entirely through non-covalent interactions, leads to a significant increase in viscosity, which is apparent immediately in the ampoule. Furthermore we have recently shown that individual fibrils further associate, presumably through secondary interactions, providing a hierarchy of supramolecular structures some of which are large enough to be easily visible in the optical microscope.[5] The mechanisms and thermodynamics of amyloid fibril self-assembly, and the morphology of the viscous solutions formed, still remain largely unclear.

DOI: 10.1002/masy.200550412

Such knowledge of the structure-property relationships could facilitate the design and implementation of novel biomaterials based on mis-folded proteins.

Bovine insulin is an ideal model to use to study the self-assembly of amyloid fibrils due to it's non-toxicity and its physiological importance, as it can form fibrils both in vivo and in vitro. Insulin is a small, helical protein consisting of two polypeptide chains, a 21 residue long A chain and a 30 residue long B chain, linked together by two disulphide bridges and in vitro it readily self-assembles following incubation in solution at pH 2.0 and temperatures above 30 °C. Under such conditions it forms viscous solutions that contain beta-sheet fibrils with an average diameter of ca. 10 nm. Here we focus on the formation of the hydrogel and the influence of environmental conditions, such as gelation temperature and protein concentration, on the kinetics of gelation and morphology of the resulting gel.

Materials and Methods

Samples. Bovine pancreatic insulin and all other chemicals were purchased from Sigma-Aldrich Company (Sigma-Aldrich co., Gillingham, UK) and used without purification. Solutions were made up by dissolving the protein powder in doubly distilled water and the pH adjusted with HCl to pH 2. Samples were incubated at temperatures ranging from 55 to 8-°C for 24 hours or until a gel had formed.

Environmental Scanning Electron Microscopy (ESEM). For all ESEM experiments reported here, we used a FEI Quanta 200 instrument with a Peltier stage and gaseous secondary electron detector (GSED) to produce an image. All samples were contained within a circular brass stub with a diameter of 1 cm and a depth of 0.5 cm. The temperature of the stub, and hence sample, was controlled using the Peltier cooling stage with a water/propanol coolant maintained at 283 K. The sample was placed on the stub, left for a few minutes to equilibrate, before initiating pump down of the chamber from ambient to a few torr. In all cases, the sample was cooled to 275 K (2 °C) and surrounded by water vapour at a pressure of ca. 4 torr. Working distances were typically 8-10 mm and images were collected using an accelerating voltage of 5 keV. Under such conditions signal to noise ratio was maximized, sample dehydration minimized, and detailed high-resolution images were obtained. Dimensions of the fibrillar network and mesh sizes were subsequently determined directly from the images using computer software built into the ESEM.

Micro Differential Scanning Calorimetry (µDSC). Highly sensitive DSC measurements (Seteram) were carried out by filling the sample cell with ca. 0.5 ml solution, while the reference cell was filled with a matching buffer. All samples were equilibrated at 10 °C for 1 hour prior to any experiment. A scan rate of 0.5 °C min^{-1} was selected for all samples, and isothermal temperatures were varied to include: 55, 60, 65, 70, 75 and 80 °C.

Results and Discussion

The morphology of the viscous solutions (hydrogels) formed after exposing insulin solutions to heat was monitored using environmental scanning electron microscopy (ESEM). Typically gels formed after 4-6 hours, and if left for prolonged periods of time particulate structures, reminiscent of spherulites, begin to form and drop out of solution.[5] A typical micrograph of insulin hydrogel is given in Figure 1a where the network is composed of entangled macrofibrils and pores ranging in size from 4 to 10 μm. In all cases, the macrofibrils forming the networks are 1–2 *μm* in diameter and are of indefinite length. Slowly dehydrating the sample inside the ESEM chamber (Figure 1b) revealed that the macrofibrils are made up of what appear to be discs that are stacked face to face. These discs have a diameter equal to the length of one amyloid fibril, suggesting that they are composed of self-assembled beta-sheet fibrils, but the exact orientation of these fibrils in the discs is still not known.

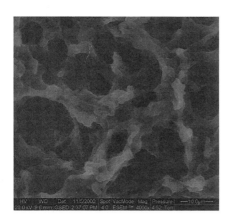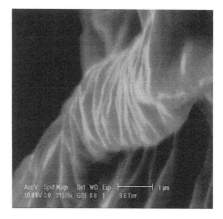

Figure 1. Typical ESEM micrographs of an 11 mgml^{-1}, pH 2, hydrogel sample obtained from aggregation of fibrillar insulin protein (6 hours, 75°C), where sample in a) is hydrated, but sample b) has been dried in order to obtain further structural detail. Scale bar represents a) 10 and b) 1μm.

Incubation temperature was found to have little effect on macrofibril formation or its dimensions, but temperature did affect the number of fibrillar entanglements and consequently mesh size: the higher incubation temperature, the greater the number of entanglements and consequently the smaller the hydrogel mesh size (Figure 2). A similar decrease in mesh size was observed when the concentration of protein was increased. To investigate such differences further sensitive micro differential scanning calorimetry (μDSC) was employed to identify the denaturation temperature and to detect the onset, kinetics and enthalpy of gelation.

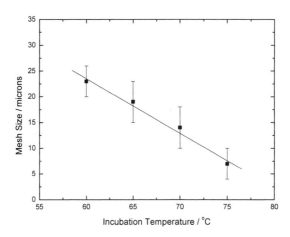

Figure 2. Average mesh size of the fibrillar network as a function of incubation temperature.

Figure 3 shows the DSC scan obtained for an 11 mg.ml[-1] insulin sample, before and after incubation, i.e. gelation. Such data are typical of all samples run here. During the initial heating ramp an endothermic peak with a minimum value at 57.5 ± 1 °C was observed due to denaturation of the protein. The heat capacity of this transition was always ca. 2 kJ mol[-1], which is typical for insulin.[6] The DSC scan of the sample after incubation showed no endothermic peak, implying that all protein had denatured and had been incorporated into the fibrillar network during the first heating ramp, and moreover that this gelation process is not reversible.

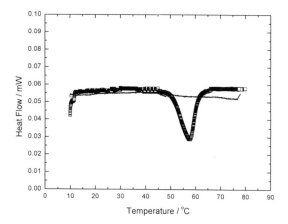

Figure 3. DSC scan of the endothermic denaturation peak of 11 mg.ml⁻¹, pH 2, insulin sample before (□) and after ($\cdots$) incubation. Average scan rate was 0.5 °Cmin⁻¹.

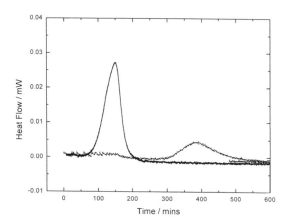

Figure 4. Isothermal calorimetry curve for 11 mgml⁻¹, pH 2 insulin sample incubated at 75 °C (—) and 60 °C ($\cdots$).

To determine the thermodynamics of protein gelation isothermal calorimetric scans at 6 temperatures were recorded for 6 hours. At 55 °C no peak was observed in the DSC trace, presumably due to the protein not having denatured and consequently not able to aggregate into beta-sheet rich fibrils. Alternatively gelation could have been too slow to be detected. At

60 °C a distinct, broad exothermic peak was observed (see Figure 4) after 314 ± 12 minutes. As the incubation temperature increased the exothermic peak began to appear more quickly and became narrower in shape (Figure 4), suggesting that protein aggregation, and consequently gelation, occurred more rapidly. Furthermore as the gelation rate increased, so too did the exothermic character (Table 1); for example gelation began 314 ± 12 minutes after the incubation time began for the sample at 60 °C with 65 ± 3 kJ mol^{-1} energy, but only after 68 ± 8 minutes and 131 ± 4 kJ mol^{-1} for the sample incubated at 75 °C. Such an increase in enthalpy of gelation suggests that the degree of gelation, i.e. number of entanglements, has also increased. This is confirmed from ESEM experiments where the number of macrofibril entanglements increased, and the average mesh size decreased with increasing incubation temperature (Figure 2). At 80 °C no well-defined exothermic peak was detected as the protein had already formed a gel before the isothermal temperature had been reached.

Table 1. Heat of gelation for 11 mgml^{-1}, pH 2 aggregated insulin fibrils incubated at different temperatures.

Temperature / °C	ΔH / kJ mol^{-1}	Time taken to reach maximum of peak / mins
55	-	No peak observed
60	65 ± 3	314 ± 12
65	79 ± 4	233 ± 10
70	114 ± 6	108 ± 5
75	131 ± 4	68 ± 8
80	Not able to determine	24 ± 10

Conclusions

The results of the present study show that under conditions that favour insulin denaturation the protein self-assembles to form beta-sheet rich fibrils. These consequently aggregate further into macrofibrils that become entangled and result in the formation of a fibrillar hydrogel. The rate and degree of macrofibril entanglements, and consequently hydrogel mesh size, can be manipulated by varying protein concentration and incubation temperature.

[1] C. M. Dobson, *Phil. Trans. R. Soc. Lond. B* **2001**, *356*, 133.
[2] C. M. Dobson, *TIBS*, **1999**, *24*, 329.
[3] C. E. MacPhee, C. M. Dobson, *J. Amer. Chem. Soc.* **2000**, *122*, 12707.
[4] J. W. Kelly, *Curr. Opin. Struct. Biol.* **1998**, *8*, 101.
[5] A. F. Miller, A. M. Donald, C. M. Dobson, C. E. MacPhee, *Accepted by Proc. Natl. Acad. Sci.U.S.A.Sept.* **2004**.
[6] W. Dzwolak, R. Ravindra, J. Lendermann, R. Winter, *Biochemistry* **2003**, *42*, 11347.

Structural Evolution Process in Solvent-Induced Crystallization Phenomenon of Syndiotactic Polystyrene

Kohji Tashiro

Department of Macromolecular Science, Graduate School of Science, Osaka University, Toyonaka, Osaka 560-0043, Japan

Summary: Structural change has been traced in the solvent-induced crystallization phenomenon of syndiotactic polystyrene through the time-resolved measurements of infrared and Raman spectra and X-ray diffraction. Immediately after the solvent is supplied to the glassy sample, the random coils start a micro-Brownian motion and locally change to short regular helical segments after some induction time. These segments grow longer and gather together to form the crystal lattice. This crystallization occurs even at room temperature far below the original glass transition temperature (Tg = ca. 100°C), because Tg is shifted to ca. -90°C (in the case of chloroform) due to the plasticizing effect, as revealed by the temperature-dependent infrared spectral measurement and the molecular dynamics calculation. The thus-created sPS-solvent complex was found to show a fast and reversible solvent exchange phenomenon between the originally-existing solvent (toluene, for example) and the newly-supplied different type of solvent (chloroform, for example). The time-dependent measurement of wide-angle and small-angle X-ray scatterings using a synchrotron radiation source revealed that the solvent exchange occurs with keeping both the columnar structure of the crystal and the stacked lamellar structure, and that the solvent exchange rate is in the order of chloroform > benzene > toluene, reflecting the difference in diffusion rate of solvent molecules and polymer-solvent interaction.

Keywords: crystallization; infrared spectra; solvent exchange phenomenon; synchrotron X-ray scattering; syndiotactic polystyrene

Introduction

syndiotactic Polystyrene (sPS) is known to form a complex (δ form) with such solvent molecules as toluene, chloroform etc. when the glassy sample is exposed to a solvent atmosphere at room temperature. The molecular chains take a regular helical conformation of TTGG type (T: trans and G: gauche). These helical chains are surrounded by solvent molecules to form a columnar structure as shown in Figure 1.[1] This columnar

 DOI: 10.1002/masy.200550413

structure is unique in a point that the originally-existing solvent molecules can be exchanged easily with newly-supplied different type of solvent molecules.[2] In order to clarify the formation mechanism of sPS-solvent complex and the fast solvent exchange phenomenon of the complex, the time-resolved measurements of infrared and Raman spectra and wide-angle and small-angle X-ray scatterings were made in these processes. The quantitative analysis of these experimental data in combination with the computer simulation allowed us to make a concrete imagination about the structural changes occurring in these phenomena as functions of time.[2-7]

Structural Evolution in Solvent-induced Crystallization Process

The glassy sample was set into a cell, into which the solvent vapor was supplied at a predetermined constant temperature. Figure 2 shows the infrared spectral changes observed at 20 and 9°C for the case of toluene as a solvent. The absorbances of infrared bands characteristic of helical conformation are plotted against time as shown in Figure 3, where the amorphous band is also plotted for comparison. As reported already the infrared (and Raman) bands can be distinguished depending on the sensitivity to the regular helical segmental length or the so-called critical sequence length (m).[3, 4] In Figure 3 the band at 549 cm^{-1} has an m value of 7 – 12 monomeric units and it was detected for the first time around 6 min after the solvent supply. This means a start of formation of short helical segments consisting of 7 – 12 monomeric units or 2 – 3 helical turns. The band at

(a)   (b)   (c)

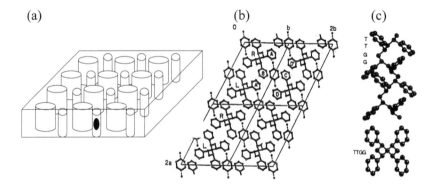

Figure 1. (a) An illustration of sPS-solvent complex crystal. Large cylinders indicate the sPS helical chains shown in (b)[1] and (c). Thin cylinders indicate the columns of solvent.

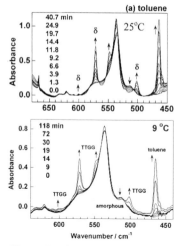

Figure 2. Time dependence of infrared spectra of sPS glass observed in the course of supplying toluene vapor at 25 and 9°C[7].

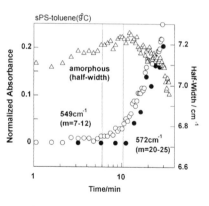

Figure 3. Time dependence of the infrared absorbances estimated for two bands with longer and shorter critical sequence length m and the half width of the amorphous band[7].

572 cm^{-1} corresponding to longer helical length ($m = 20 - 25$ monomeric units or 5-6 helical turns) was observed later. The difference in time to detect these two bands indicates a growth of short helical segment during this time lag. At the same time it is noticed that the half width of the amorphous band increased rapidly after the supply of solvent. Since the band width is inversely related to the molecular mobility as reported in the previous paper,[4] this increase reflects an activation of molecular motion in the amorphous region. Therefore, the data given in Figure 3 tells us the following structural regularization process as illustrated in Figure 4[7]. (The essentially same picture was presented in a previous paper based on the experimental data at room temperature,[4] but the data taken at 9°C are useful for detecting the structural evolution process more clearly because the

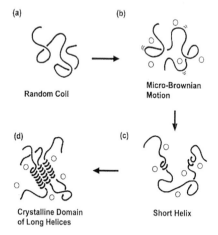

Figure 4. Structural evolution of sPS random coils in the solvent-induced crystallization process [4].

change proceeds more slowly.) (i) Immediately after the supply of solvent, the micro-Brownian motion in the amorphous region is enhanced as known from the increment of half width of the amorphous band. (ii) As a result, the short but regular helical segments are generated and they grow gradually to longer helices. (iii) These helical segments gather together to form a crystalline lattice, as revealed by the X-ray diffraction data[3, 4]. (iv) The measurement of small-angle X-ray scattering revealed the formation of higher order structure with a long period of ca. 100 Å. That is to say, the crystalline domains shown in Figure 4 (d) are stacked along the chain axis with a period of ca. 100 Å.

As seen in Figure 2, a growing rate of crystalline infrared bands decreases remarkably as the crystallization temperature is lower, and the finally-attained crystallinity is relatively low. The time dependence of infrared absorbance was analyzed on the basis of Avrami equation (Absorbance $D = D_0*\exp[-k(t - t_0)^n]$, where k is a rate constant, t_0 is an induction period and n is an index of crystallization). The rate constant k was plotted against temperature. An extrapolation to $k = 0$ gives the temperature below which solvent-induced crystallization does not occur anymore, or the apparent Tg under an existence of solvent. Similar analysis was made also for the crystallinity vs temperature curve. The thus-estimated Tg was -90 ± 10 °C for chloroform, -70 ± 10°C for benzene, and -30 ± 10°C for toluene[6, 7]. In this way the plasticizing effect is remarkable and the Tg is shifted by 130 – 190°C depending on the type of solvent. It is understood why the glassy sPS sample with originally high Tg can be crystallized even at room temperature when the solvent is supplied to the system. This plasticizing effect could be simulated reasonably by performing the molecular dynamics calculation for the model consisting of sPS amorphous chains and solvent molecules at some molar ratios. As shown in Figure 5, a deflection point of the volume or Tg is shifted to lower temperature side depending on the sPS/solvent ratio, where toluene is used as solvent, although the absolute Tg value is rather higher than the observed ones[6, 7].

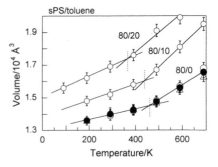

Figure 5. Temperature dependence of volume of sPS/toluene system calculated by molecular dynamics method for the various sPS/toluene molar ratio [7].

Solvent Exchange Phenomena

The thus created sPS-solvent complex was found to show such a characteristic behavior that the solvent molecules originally contained in the complex are easily exchanged with the newly-supplied different type of solvent. In the previous paper [2] we measured the time-dependence of infrared spectra in the course of solvent exchange process between a pair of solvents (chloroform, benzene, and toluene). Recently we have performed the time-resolved measurements of small-angle (SAXS) and wide-angle X-ray scatterings (WAXS) using synchrotron radiation systems (Photon Factory of KEK and Spring-8) for both the uniaxially-oriented and unoriented samples. The as-drawn sample showed the X-ray fiber pattern of planar-zigzag chain conformation and no long period was observed along the draw axis (Figure 6). When this sample was exposed to toluene atmosphere, it changed to the δ form and showed clear meridional SAXS peaks corresponding to the long period of ca. 100 Å. Figure 7 (a) shows the time dependence of meridional SAXS profile measured for the oriented sPS-toluene complex. When chloroform was supplied to this sample, the SAXS peak decreased in intensity due to the X-ray absorption effect by chlorine atoms, during which the WAXS pattern changed correspondingly with keeping the orientation of crystallites unchanged. On supplying benzene to this chloroform-containing sample, the exchange from chloroform to benzene occurred and the SAXS peak increased the intensity

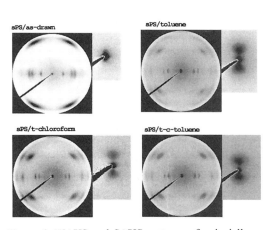

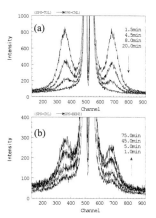

Figure 6. WAXS and SAXS patterns of uniaxially oriented sPS samples. (a) as-drawn sample, (b) exposing (a) into toluene, (c) the sample (b) exposed in chloroform, and (d) the sample exposed again in toluene.

Figure 7. Time dependence of SAXS profile of sPS-solvent complex measured during the solvent exchange process from (a) toluene to chloroform and (b) chloroform to benzene.

again [Figure 7 (b)]. Figure 8 shows the SAXS intensity change during these phenomena observed for an unoriented sample. The solvent-induced crystallization occurred at low rate. But, once the complex was formed, the solvent exchange occurred at higher rate. Among the three solvent exchange processes, the exchange was found to occur in the order of chloroform > benzene > toluene. (In the previous paper[2] we described the infrared spectral data and suggested almost the same solvent exchange rate between the three cases. However, this suggestion must be corrected, and the exchange rate should be different as indicated in Figure 8. In the present experiment of WAXS and SAXS measurements, a thicker sample was used and the crystallization and solvent-exchange phenomena could be traced at longer time scale, giving more confirmative result than the faster

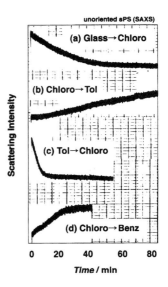

Figure 8. Time-dependence of SAXS intensity measured in crystallization and solvent-exchange processes of an unoriented sPS sample.

infrared experiment using a thinner film.) The order of solvent exchange rate is almost parallel to that of the crystallization rate. The solvent-induced crystallization rate is affected by the solubility of sPS in a solvent and the diffusion rate of solvent molecules in the amorphous phase. Chloroform molecule is small in size and dissolves the sPS more easily than benzene or toluene. Such solvent effect is considered to govern also the solvent exchange rate.

[1] Y. Chatani, Chatani Y, Shimane Y, Inagaki T, Ijitsu T, Yukinari T, Shikuma H. Polymer 1993; 34: 1620.
[2] A. Yoshioka, K. Tashiro, Macromolecules 2003, 36, 3593.
[3] K. Tashiro, Y. Ueno, A. Yoshioka, M. Kobayashi, Macromolecules 2001, 34, 310.
[4] K. Tashiro, A. Yoshioka, Macromolecules 2002, 35, 410.
[5] A. Yoshioka, K. Tashiro, Macromolecules 2003, 36, 3001.
[6] A. Yoshioka, K. Tashiro, Polymer 2003, 44, 66811.
[7] A. Yoshioka, K. Tashiro, Macromolecules 2004, 37, 467.

Temperature-Concentration Phase Diagram of PEO-Urea

Jean-François Wagner,[1] *Marcel Dosière,*[2] *Jean-Michel Guenet*[*1,2]

[1.]Institut Charles Sadron, CNRS UPR22, 6 rue Boussingault F-67083 Strasbourg Cedex, France
E-mail: guenet@ics.u-strasbg.fr
[2.]Laboratoire de Physico-chimie des Polymères, Université de Mons-Hainault, Place du Parc, 20 B-7000 Mons, Belgium

Summary: The present paper reports on the phase diagram of the system PEO/Urea as established by differential scanning calorimetry. From this phase diagram one PEO/Urea compound is indentified with a stoichiometry 1/2. This value is in agreement with results derived from X-ray diffraction.

Keywords: compounds; phase diagram; poly[ethylene oxide]; urea

Introduction

Poly[oxyethylene] (PEO) forms compounds with a large variety or organic and inorganic molecules, such as urea and thiourea.[1-4] The phase diagrams in many systems, particularly with *para*-dihalogeno-benzene molecules,[1] have been reported. From these diagrams the number and the stoichiometry of the compounds have been obtained. While many investigations have been carried out with urea, which was the first PEO/solvent compound ever observed, the temperature-concentration phase diagram is still missing. This paper presents this phase diagram as obtained from differential scanning calorimetry. The results are discussed in light of the molecular structure derived from X-ray diffraction experiments.[5]

Experimental

The PEO samples were purchased from Aldrich. Two samples were used: one with a weight-averaged molecular weight of $M_w = 3 \ 10^5$ g/mol with and polydispersity $M_w/M_n = 2$ for the high urea contents, and one $M_w = 6 \ 10^3$ g/mol for PEO contents above $C_{PEO} = 0.8$ (w/w). Urea was also purchased from Aldrich and use without further purification.

The PEO/urea samples were first prepared in test tubes by heating above the melting temperature of urea and by applying a vigorous stirring in order to obtain homogeneous solutions. They were then cooled down to room temperature. About 10 mg were taken out

© 2005 WILEY-VCH Verlag GmbH & KGaA, Weinheim

DOI: 10.1002/masy.200550414

from the test tubes, and then placed into DSC sample pans that were hermetically sealed. Before taking any data, the samples were melted well above the melting temperature of urea and of the compound (namely at 160°C) and cooled down to room temperature. Differential calorimetry measurements were performed on a DSC4 from PERKIN-ELMER at a heating rate of 5°C/min.

The optical micrographs were obtained on a NIKON OPTIPHOT-2 equipped with a digital camera DXm1200, and using LCIA software for image processing and analysis.

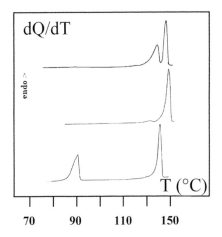

Figure 1. Typical thermograms obtained by differential scanning calorimetry at a heating rate of 5°C/m. Upper C_{PEO}= 0.1 (w/w), middle C_{PEO}= 0.25 (w/w), lower C_{PEO}= 0.6 (w/w).

Results and Discussion

As can be seen in figure 1, three types of endotherms can be observed. For *low PEO* contents two endotherms are seen at high temperature, that are close to the urea melting point. Close to what will be identified as the stoichiometric composition of the PEO/urea compound, only one endotherm is essentially observed, while for *high PEO* contents a second endotherm appears at lower temperature, that turns out to be actually close to pure PEO melting temperature.

The corresponding temperature-concentration phase diagram is plotted in figure 2, together with the corresponding Tamman's diagram (latent heats ΔH associated with all the 1st order thermal events as a function of the polymer concentration in w/w).

This phase diagram clearly evidences the formation of a PEO/urea complex whose stoichiometric composition is $C= 0.26\pm0.03$, and which displays congruent melting. The stoichiometric composition is derived both from the temperature-concentration phase diagram and from the Tamman's diagram. As a matter of fact, the enthalpy associated with the terminal melting goes, as expected, through a maximum at this composition. Also, the enthalpies associated with the eutectic events (invariance of temperatures at $T= 135°C$ for $C< 0.26$ and $T= 60°C$ for $C> 0.26$) become zero at the stoichiometric composition. The value $C= 0.26$ yields a stoichiometry of about *1 PEO monomer/2 urea molecules*. This value is in excellent agreement with what has been computed from the cristalline lattice derived from X-ray data [5].

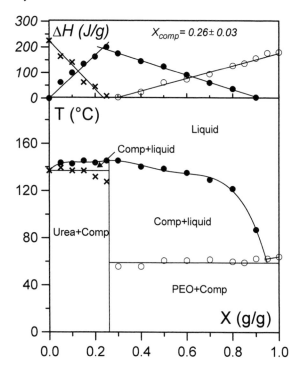

Figure 2. Temperature-concentration phase diagram PEO/Urea. On top Tamman's diagram representing the melting enthalpies ΔH (Joules per gramme of material) associated with the different thermal events (symbols correspond in both diagrams).

The morphology of the compound is shown in figure 3 from an optical micrograph obtained after quenching to room temperature a PEO/Urea sample whose concentration is

124

close to the stoichiometric composition. As can be seen large spherulitic stuctures are obtained. Note that no such structures exist for the pure urea, unlike what is seen for pure PEO.

Figure 3. Optical micrograph obtained from a 0.3 g/g PEO/Urea, namely close to the stoichiometric composition.

Concluding remarks

The temperature-concentration phase diagram PEO/Urea is now available, and can thus be compared to previous diagrams obtained for PEO/dihalogenobenzenes. The compound formed with urea is a congruently-melting compound as with most of the PEO/dihalogenobenzenes. The stoichiometric with urea reaches, however, its highest value ever found for PEO complexes.

It is worth emphasizing that the stoichiometry is straightforwardly obtained from the Tamman's diagram, namely from the variation of the enthalpies associated with the different thermal events. This again emphasizes that this type of diagram is highly relevant [6], and thus very helpful for determining the stoichiometric composition with great accuracy.

Acknowledgements: J.M. Guenet is greatly indebted to the FNRS (Fond National de la Recherche Scientifique, Belgium) for a Professorship grant at Mons-Hainaut University in the academic year 2003-2004.

[1] M. Dosiere, *Macromol. Symp.* **1997**, *114,* 63
[2] J. Parrod, A. Kohler, G.Hild, *Compt. Rend. Acad. Sci.* **1958**, *246,* 1046
[3] J. Parrod, A. Kohler, G. Hild, *Makromol. Chem.* **1964**, *75,* 52
[4] H. Tadokoro, T. Yoshihara, Y. Chatani, S. Murahashi, *J. Polym. Sci.* **1964**, *B2,* 363
[5] A. Chenite, F. Brisse, *Macromolecules* **1991**, *24,* 2221
[6] J.M. Guenet, *Macromol. Symp.* **2003**, *203* 1

Macromol. Symp. **2005**, *222*, 125-133

Structure and Properties of Intramolecular Polycomplexes Formed in Graft Copolymers with Chemically Complementary Polymer Components

Tatyana Zheltonozhskaya

Macromolecular Chemistry Department, Kiev Taras Shevchenko National University, 64 Vladimirskaya Str., 01033, Kiev, Ukraine
E-mail: zheltonozhskaya@ukr.net

Summary: The structure in a bulk state and in a solution and also peculiarities of synthesis of the polyacrylamide to poly(vinyl alcohol) graft copolymers forming the intramolecular polycomplexes (IntraPC) are reviewed basing on the data of IR spectroscopy, dilatometry, DSC, QELS, static light scattering, rheology and some other methods. Influence of the quantity N and molecular weight of grafted chains and also the hydrogen bond system between the main chain and the grafts and between the grafted chains only on the IntraPC properties is highlighted. As illustration of the high binding ability of IntraPCs the results of studying of their complex formation with silica nanoparticles in aqueous medium are discussed.

Keywords: binding ability; graft copolymer; hydrogen bonds; intramolecular polycomplex; polymer-colloid complex; structure

Introduction

Binding properties of intermolecular polycomplexes (InterPC) with respect to low- and high-molecular-weight organic substances (first of all drugs and biopolymers), different ions, colloid particles and cells of living organisms are well known.[1-3] But as binders they have essential demerits conditioned by existence of the complex formation equilibrium at the macromolecular level. Really, InterPC can be irreversibly are destroyed up to individual macromolecules at a considerable dilution of a solution and pH alteration (if one or both polymer components are polyelectrolytes) and also at the action of more strong competitors or such factors as a temperature and additives of the inert low-molecular weight electrolytes.[1-3] In this connection the intramolecular polycomplexes (IntraPC) formed in macromolecules of block and graft copolymers by chemically complementary components are of special interest. Destruction of the cooperative system of bonds in such IntraPC is always reversible, that is why they have abnormally high binding ability.

DOI: 10.1002/masy.200550415

Peculiarities of a synthesis and structure in a bulk state and in a solution as well as factors of stabilization of IntraPCs formed in the polyacrylamide to poly(vinyl alcohol) graft copolymers (PVA-*g*-PAA$_N$, where N is the grafted chain number) are considered in the present paper. Their binding ability in respect of silica nanoparticles is demonstrated.

Properties and Binding Ability of Intramolecular Polycomplexes

Synthesis. Three series of PVA-*g*-PAA$_N$ copolymers were synthesized by free radical graft copolymerization with a Ce (IV) salt as the initiator according to Ref.[4] In a series of the PVA-*g*-PAA$_N$1-3 samples N was varied from 25 to 49 but molecular weights of the grafts were comparable ($M_v\sim1\cdot10^5$).[5] In another series of the PVA-*g*-PAA$_N$4-6 copolymers N was changed from 25 to 42 and the molecular weight of the grafts was decreased from $3.72\cdot10^5$ to $1.63\cdot10^5$.[6] Then a series of PVA-*g*-PAA$_N$7-9 samples was prepared in which M_{vPAA} increased from $3.72\cdot10^5$ to $5.1\cdot10^5$ at the constant N=9.[6] Until recently the mechanism of grafting of PAA onto hydroxyl-containing polymers (including PVA) was believed to be well known.[8] But basing on thermodynamic affinity between PVA and PAA and also on the phenomenon of the matrix polymerization of monomers of the one chemical nature in the presence of complementary macromolecules of another chemical nature,[9] the idea about a matrix character of the PAA to PVA graft copolymerization has appeared. This assumption was examined in the kinetic studies of synthesis of the PVA-*g*-PAA$_N$1-3 and PVA-*g*-PAA$_N$7-9 copolymers, which were performed by dilatometry.[12] The rate of the graft copolymerization was shown in 2-4 times higher than that one of the acrylamide homopolymerization carried out in the same experimental conditions. Not only the kinetic matrix effect has been established in the PVA-*g*-PAA$_N$ synthesis, but also its relation to the density of the growing chains has been revealed. Really, when the quantity (density) of the growing grafts became too high (as in synthesis of PVA-*g*-PAA$_N$3) the kinetic matrix effect disappeared).[12]

The H-bond System Stabilizing IntraPC. Studying of the system of intramolecular bonds depending on the number (density) and molecular weight (length) of grafts is of fundamental importance for graft copolymers forming IntraPC because such system shapes already in the process of their synthesis (matrix effect). Hence, considerable attention has been given to this problem in researches of the PVA-*g*-PAA$_N$ graft copolymers. Careful quantitative analysis of the IR spectroscopy data basing on the computer separation of overlapping vibration bands in the Amide I and Amide II regions by the spline method was carried out. It was shown that IntraPC structure in the PVA-g-PAA$_N$ macromolecules is stabilized by both H-bonds between the main chain and the grafts and also H-bonds between the units of the grafted chains (such as *cis-tras*-multimers of amide groups).[6,13,14] An increase in N in the series of PVA-g-PAA$_N$1-3 led to substantial changes in this H-bond system (Figure 1).

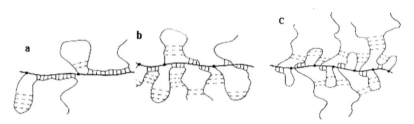

Figure 1. Redistribution of hydrogen bonds in the PVA-g-PAA$_N$ copolymers depending on N.

Actually, the number of H-bonds such as the main chain – the grafts decreased with N growth whereas the number of H-bonds between the grafts increased.[14] At the same time, interesting effect of 'detachment' and stretching of the grafted chains in different directions away from the main chain was observed for the series of PVA-g-PAA$_N$7-9 copolymers when molecular weight (length) of the grafts became higher than some critical value M* (Figure 2): [6,13,14]

The changes in the hydrogen bond system with increasing N and M_{vPAA} were reflected in all the properties of PVA-g-PAA$_N$.

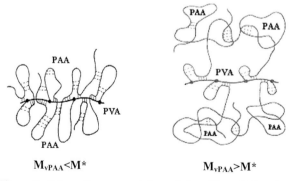

$M_{vPAA}<M*$ $M_{vPAA}>M*$

Figure 2. Schematic representation of the effect of 'detachment' of the grafts from the main chain.

Influence of Grafting Density. The bulk structures of the PVA-*g*-PAA$_N$1-3 graft copolymers with variable N were studied by DSC method.[15] It was found that the homogeneous amorphous copolymer structure characterized by alone glass transition remained up to some critical value of N (N*~25). But at N>N* two glass transitions on the DSC thermograms appeared, that was conditioned by a microphase separation in the bulk structure of the graft copolymers. In this case side by side with the regions of full compatibility of the polymer components domains formed by the PAA segments, which do not interacted with the main chain because of steric hindrances, arise.

According to the data of static light scattering[5], water was practically the Θ-solvent in respect of the copolymers, but its thermodynamic quality was essentially improved with N growth. Such behavior of the graft copolymers in solution is consistent with the effect of a successive decrease in the number of the main chain-the grafts H-bonds in IntraPC. The contribution of hydrophobic interactions of bound parts of active groups of the polymer components to stabilization of IntraPC structures in water was estimated by benzene solubilization [14]. The results confirmed that such stabilization occurs and the role of this factor substantially decreases with increasing N. Actually, the volume of hydrophobic regions in the PVA-*g*-PAA$_N$ 1-3 series reduced from 794 to 98 nm^3.

Investigations of the changes in the state of the PVA-*g*-PAA$_N$1-3 and 4-6 macromolecules in solutions induced by factors capable to destroy the H-bond system in IntraPC were of special interest. Among these factors, the temperature[16] and hydrodynamic shear field[5] were examined. It was shown by viscometry and quasielastic laser scattering [16] that the copolymer macromolecules undergone the reversible conformational transition such as *IntraPC↔ segregated state* in the narrow temperature interval from 303 K to 308 K. Destruction of the H-bond system in IntraPC and essentially different solubility of PVA and PAA in water were the main reasons of such transition. The effect of hydrodynamic shear flow on the copolymer structure in solution was studied on a rheo-viscometer with coaxial cylinders [5]. Two levels of the copolymer reversible structure destruction, which were attributed to: i) destroying of a network of pinnings only (first level) at the shear rate gradient j lesser than some critical value j* and at the copolymer concentration C higher than C* and also ii) ruining of both the Intra PCs and their primary aggregates (second level) at j>j* and C< C*, were revealed. Stability of PVA-*g*-PAA$_N$ structure to a hydrodynamic shear field reduced when N increased.

Influence of the Molecular Weight (Length) of Grafts. When studying the structure and behavior in solutions of the PVA-*g*-PAA$_N$7-9 samples with constant N but different molecular weight of the grafts the main interest was to elucidate how the effect of 'detachment' of the grafts from the main chain (which was revealed by IR spectroscopy) is displayed in the copolymer properties.[13,14,17] DSC researches confirmed the phenomenon of 'detachment' of the grafts by discovering of the state of microphase separation (two glass transitions in DSC thermograms) in the copolymer bulk structure at $M_{vPAA} > M^*$ in contrast with the homogeneous copolymer structure at $M_{vPAA} \leq M^*$.[17] Due to above-mentioned phenomenon, macrocoils of the PVA-*g*-PAA$_N$9 sample having M_{vPAA} higher than M^*, sharply swelled in aqueous solutions (comparing with the PVA-*g*-PAA$_N$7-8 samples) and formed the developed friable aggregates.[13] Additionally, basing on the investigations of benzene solubilization, the conclusion, about destroying of hydrophobic regions and a sharp increase in the amount of H-bond benzene in swollen coils of the PVA-*g*-PAA$_N$9 sample, was made.[18]

Binding Properties of the Intramolecular Polycomplexes. Graft copolymers forming IntraPC, such as PVA-*g*-PAA$_N$ are binders of new generation. Their particles in aqueous medium contain the hydrophobic regions and hydrophilic cavities formed by loops of non-bound polymer units (including active polymer groups), therefore they have practically universal binding ability. In particular, flocculants based on these graft copolymers can efficiently clear a river water on waterworks from pollutants of the natural and anthropic origin.[19,20] The interaction of PVA-*g*-PAA$_N$ with humic and fulvic acids (the main components of the humus substances in a river water) and also with phenol (one of the toxic substances in water) and the ions of stable and radioactive isotopes of cesium and strontium was considered earlier.[21,22] At the same time studying of the copolymer complex formation with smallest colloid particles are of great interest in order to show their binding possibilities. Two PVA-*g*-PAA$_N$2 and 3 samples were used to investigate their interaction with silica sol ($R_{SiO2}=11,2$ nm), which was prepared from Aerosol A-175 ('Oriana', Ukraine), by viscometry, static light scattering and benzene solubilization. The polymer solutions were mixed with SiO_2 sol during 1 hour in the rations corresponding to 2 and 5 nanoparticles on every macromolecule. Formation of the polymer-colloid complexes (PCC) was accompanied by a sharp decrease of the reduced viscosity of the copolymer solutions (Figure 3). When the number of SiO_2 particles was higher the reduced viscosity decreasing ($\Delta\eta_{red}$) was greater. The PCCs appeared was stable in aqueous

130

medium. In particular, they displayed greater stability to dilution of a solution than the initial IntraPCs (Figure 3, curves 2,3).

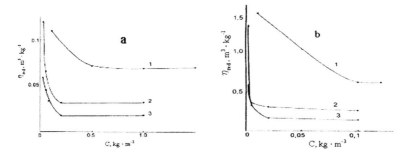

Figure 3. The reduced viscosity *vs* concentration for PVA-*g*-PAA$_N$2 (a) –*1*, PVA-*g*-PAA$_N$3 (b) –*1* and also for their polymer-colloid complexes with two –*2* and five –*3* SiO$_2$ particles.

The depolarization coefficients Δ_v and Δ_u (for vertically polarized and unpolarized falling light), which are sensitive to the smallest structural changes in macrocoils [23] were measured by instrument of the static light scattering FPS-3 (Russia) in the solutions of PVA-*g*-PAA$_N$2,3 and their mixtures with colloid nanoparticles (Figure 4).

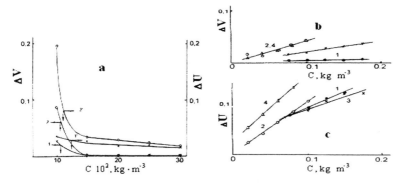

Figure 4. Depolarization coefficients *vs* concentration for PVA-*g*-PAA$_N$2 -*1,1'* (a), PVA-*g*-PAA$_N$3 -*2,2'* (a) and also for the mixtures PVA-*g*-PAA$_N$2+SiO$_2$ –*1,2* (b,c) and PVA-*g*-PAA$_N$3+SiO$_2$ –*3,4* (b,c) with two (*1,3*) and five (*2,4*) SiO$_2$ particles. Θ=90°, λ=436 nm.

It is seen that the IntraPC structures in the graft copolymers are keeping into a wide region of their concentrations (Figure 4, a). However at C < 0,15 kg·m^{-3} the sharp macrocoil swelling due to the IntraPC destruction occurs. Unlike to behavior of individual graft copolymers, their PCCs with 2 and 5 SiO$_2$ particles demonstrated full stability in the concentration region under study. The values of $\Delta\eta_{red}$ together with Δ_v^0 and Δ_u^0 found by extrapolation of the depolarization coefficients to C=0 are represented in Table 1. The values of Δ_h^0 and δ^2 (the third depolarization coefficient and the optical anisotropy parameter) in the table were calculated according to well-known relations placed under the table.

Table 1. Characteristics of PCCs formed by the graft copolymers with SiO_2 particles.

Copolymer	$n^{1)}$	$\dfrac{\Delta\eta_{red},}{m^3 \cdot kg^{-1}}$	Δ_v^0	Δ_u^0	$\Delta_h^{0\ 2)}$	$\delta^{2\ 3)}$
	-	-	0,006	0,030	4,030	0,052
PVA-g-PAA$_N$2	2	0.54	0,003	0,032	9,70	0,055
	5	0.74	0,003	0,030	9,03	0,052
	-	-	0,020	0,049	1,499	0,087
PVA-g-PAA$_N$3	2	0.58	0,001	0,024	23,02	0,042
	5	0.77	0,001	0,027	26,30	0,046

[1)] The number of SiO_2 particles in corresponding PCC.
[2)] $\Delta_u^0 = (1 + \Delta_h^0)/ [1 + (\Delta_v^0)^{-1}]$.
[3)] $\delta^2 = 10\, \Delta_u^0/ (6 - 7\Delta_u^0)$. [23]

Note, that the depolarization coefficient Δ_v^0 is determined by the internal anisotropy of a polymer substance, whereas the depolarization coefficient Δ_h^0 depends on the anisotropy of macrocoils which is connected with their shape and large dimensions.[23] Finally, the parameter δ^2 characterizes the total anisotropy of polymer particles which is conditioned by both above-mentioned contributions.[23]

Let's consider with such point of view the data in Table 1. It is seen that the inside structure and shape of PVA-g-PAA$_N$2 macromolecules with lesser quantity (density) of the grafts (N=31) does not change enough in the complex formation with 2 and 5 SiO_2 particles. Such conclusion is based on a small alteration of the Δ_v^0, Δ_h^0 and δ^2 parameters in solution of PCCs comparing with solution of the graft copolymer. In this case the sharp decreasing of the η_{red} of the copolymer in the mixtures with SiO_2 particles can be attributed only to destroying of the PVA-g-PAA$_N$2 primary aggregates under influence of complex formation. Another situation is observed for the PVA-g-PAA$_N$3 sample with greater number of the grafts (N=49). The structure of its primary aggregates and individual macromolecules considerably change during the PCC formation. Really, the internal anisotropy and total anisotropy of macrocoils containing SiO_2 particles sharply decrease, but their size (or the coil asymmetry) grows.

The results of studying of the benzene solubilization in suspensions of individual SiO_2 particles and solutions of the PCCs (in comparison with the analogous data for the PVA-g-PAA$_N$2 and 3 samples) provide with additional information about complex formation between the graft copolymers and colloid particles. It was found that the PVA-g-PAA$_N$2 and 3 samples were able to connect in the solubilization process 3380 and 1090 mol_{benz}/mol_{cop} correspondingly. The value of adsorption of benzene by SiO_2 particles was equal to 869 $mol_{benz}/particle$. But all the polymer-colloid mixtures had no capability to solubilize benzene. Such experimental fact points on formation in these mixtures very dense PCC particles, in which the main part of active groups of both polymer components is bound with a surface of SiO_2 particles (Figure 5). It can be assume that the PVA and PAA chemical complementarity plays an important role in stabilization of the PCC structure.

Figure 5. The scheme of the PCC structure.

Conclusion

Thus, the structure and properties and also binding ability of the PVA-g-PAA$_N$ graft copolymers which belong to a class of the intramolecular polycomplexes are determined not only by the number and molecular weight of the grafts (that is typical for all graft copolymers), but also by the system of hydrogen bonds such as *a main chain-a graft* and *a graft-a graft*. Formation of the H-bond system begins evidently in the synthesis of such graft copolymers because it has a matrix character. It was shown that there are certain critical values of the number and molecular weight of the grafts which ones are dependent from each other and determine the transfer from the homogeneous copolymer bulk structure to the state of microphase separation. Due to specific building the IntraPCs can efficiently bind different low- and high-molecular-weight organic substances, some inorganic (metal) ions and also hard nanoparticles, in particular, of silica sol. It was found that IntraPC structure undergoes a significant destruction in the complex formation with SiO_2 nanoparticles if the grafting density in PVA-g-PAA$_N$ is too large.

[1] V. A. Izumrudov, A. B. Zezin, V. A. Kabanov, *Usp. Khim.* **1991**, *60*, 1570.

[2] J. Xia, P. Z. Dubin, in:*"Macromolecular Complexes in Chemistry and Biology"*, Berlin, Heidelberg: Springer **1994**, 247.

[3] V. A. Kabanov, *Vysokomolek. Soed.* **1994**, *A36*, 183.

[4] N. P. Melnik, T. B. Zheltonozhskaya, L. N. Momot, I. A. Uskov, *Dokl. Akad. Nauk Ukr. SSR*, **1988**, *B11*, 50.

[5] T. B. Zheltonozhskaya, O. V. Demchenko, N. V. Kutsevol, T. V. Vitovetskaya, V. G. Syromyatnikov, *Macromol. Symp.* **2001**, *166*, 255.

[6] V. Yu. Kudrya, V. N. Yashchuk, L. P. Paskal, T. B. Zheltonozhskaya, O. V. Demchenko, *Macromol. Symp.* **2001**, *166*, 249.

[7] S. K. Rath, R. P. Singh, *Colloids Sur.* **1998**, *A139*, 129.

[8] C. L. McCormic, L. S. Park, *J. Appl. Polym. Sci.* **1985**, *30*, 45.

[9] I .M. Papisov, *Vysokomol. Soed.***1997**, *B39*, 562.

[10] N. Permyakova, T. Zheltonozhskaya, O. Demchenko, L. Momot, S. Filipchenko, N. Zagdanskaya, V. Syromyatnikov, *Pol. J. Che.***2002**, *76*, 1347.

[11] L. N. Momot, T. B. Zheltonozhskaya, N.M. Permyakova, S. V. Fedorchuk, V. G. Syromyatnikov, *Macromol. Symp.* **2004**, *submited for publication*.

[12] N. E. Zagdanskaya, T. B. Zheltonozhskaya, V. G. Syromyatnikov, *Voprosy Khim. Khim. Tekhnol.* **2002**, *3*, 53.

[13] T. Zheltonozhskaya, O. Demchenko, I. Rakovich, J-M. Guenet, V. Syromyatnikov, *Macromol. Symp.* **2003**, *203*, 173.

[14] T. B. Zheltonozhskaya, *Doctoral Thesis in Chemical Sciences, Kyiv National University*, Kyiv, **2003**.

[15] O. V. Demchenko, N. V. Kutsevol, T. B. Zheltonozhskaya, V. G. Syromyatnikov, *Macromol. Symp.* **2001**, *166*, 117.

[16] N. Kutsevol, T. Zheltonozhskaya, N. Melnik, J-M. Guenet, V. Syromyatnikov, *Macromol. Symp.* **2003**, *203*, 201.

[17] O. Demchenko, T. Zheltonozhskaya, J-M. Guenet, S. Filipchenko, V. Syromyatnikov, *Macromol. Symp.* **2003**, *203*, 183.

[18] N. Zagdanskaya, L. Momot, T. Zheltonozhskaya, J-M. Guenet, V. Syromyatnikov, *Macromol. Symp.* **2003**, *203*, 193.

[19] Ukr. P. 23 743; 'Promislova vlastnist' **2001**, *11*.

[20] Ukr. P. 29 933; 'Promislova vlastnist' **2002**, *3*.

[21] O. V. Demchenko, T. B. Zheltonozhskaya, N. V. Kutsevol, V. G. Syromyatnikov, I. I. Rakovich, O. O. Romankevich, *Chem. Inz. Ekol.* **2001**, *8*, 463.

[22] S. Filipchenko, T. Zheltonozhskaya, O. Demchenko, T. Vitovetskaya, N. Kutsevol, L. Sadovnikov, O. Romankevich, V. Syromyatnikov, *Chem. Inz. Ekol.* **2002**, *9*, 1521.

[23] V. E .Eskin, *Ligt scattering by polymer solutions and properties of macromolecules.* Nauka, Leningrad **1986**.

Macromol. Symp. **2005**, *222*, 135-142

Structural Transitions in Triblock Copolymers Based on Poly(ethylene oxide) and Polyacrylamide under the Temperature Influence

Nataliya Permyakova,[1] *Tatyana Zheltonozhskaya,*[1] *Valery Shilov,*[1,2]
Nina Zagdanskaya,[1] *Lesya Momot,*[1] *Vladimir Syromyatnikov*[1]

[1] Taras Shevchenko National University, Chemical Department, 60 Vladimirskaya Str., 01033, Kiev, Ukraine
E-mail: permyakova@ukr.net
[2] Institute of Macromolecular Chemistry, National Academy of Sciences of Ukraine, 48, Kharkovskoe shosse, 02160, Kiev, Ukraine

Summary: Thermal transitions in the bulk structure of triblock copolymers (PAA-*b*-PEO-*b*-PAA) based on polyacrylamide and poly(ethylene oxide) with varying molecular weight (length) of PEO block comparing with the structures of individual polymers and polymer mixtures were investigated. A lot of effects, such as the melting temperature depression, decreasing of the crystallinity degree of PEO and also appearance of the microphase separation in amorphous regions of the polymer mixtures and the triblock copolymers were found. Such investigations pointed to a strong intramolecular interaction of the polymer blocks in the triblock copolymers that is confirmed by the results of IR spectroscopy. It was shown that PEO and PAA blocks formed the system of H-bonds with participant of *trans*-multimers of amide groups.

Keywords: block copolymer; crystallinity degree; hydrogen bonds; intramolecular polycomplex; microphase separation

Introduction

Graft and block copolymers with a strong interacting polymer components, which was called as intramolecular polycomplexes (IntraPC), are of great interest last years.[1] The main element of self-regulation of such systems is the thermodynamic affinity between polymer components. In present there are relatively many researches devoted to peculiarities of the bulk structures and state in solution of IntraPCs formed in the graft copolymers.[2,3] But the complex formation of chemically complementary polymers in linear block copolymers is not practically investigated. In the present work a thermal transitions in the bulk structure of triblock copolymers (TBC) based on polyacrylamide and poly(ethylene oxide) (PAA-*b*-PEO-*b*-PAA), which are capable to interact with each other were studied.

DOI: 10.1002/masy.200550416

Triblock copolymers in the bulk state

In order to establish the role of the central block length three PEO samples from "Merk" (Germany) with $M_{vPEO}=3\cdot10^3$; $4\cdot10^4$ and $1\cdot10^5$ were used in the synthesis of TBCs. Synthesis was carried out in aqueous solution according to the method of graft copolymerization of PAA onto poly(vinyl alcohol) with application of Ce (IV) salt, as the initiator.[4] The ratio between Ce (IV) and PEO concentrations was 2 $mol_{Ce\ (IV)}/mol_{PEO}$. TBC were precipitated by acetone and freeze dried. The sample of individual PAA, with $M_{PAA}=6.3\cdot10^5$, was synthesized in the same experimental conditions. TBC molecular characteristics calculated from the elemental analysis data are shown in Table 1.

Table 1. Parameters of the triblock copolymers.

Polymer	$M_{v\ PEO}\cdot10^4$	$M_{v\ PAA}\cdot10^5$	$M_{v\ PAA\text{-}b\text{-}PEO\text{-}b\text{-}PAA}\cdot10^6$	n [1]
PAA-b-PEO-b-PAA1	0.3	0.38	0.08	7.9
PAA-b-PEO-b-PAA2	4	3.18	0.68	4.9
PAA-b-PEO-b-PAA3	10	9.07	1.91	5.6

[1] The ratio of base-mol_{PAA}/base-mol_{PEO} in the copolymer macromolecules.

It is seen, that the PAA block length increase in TBC samples.

Thermal transitions in the TBC bulk structure comparing with individual polymers and mixtures of the same polymers were studied by differential scanning calorimetry (DSC) at the heating rate 16 °C/min with a Du Pont 1090 thermal analyzer. The liquid nitrogen cooled the carefully dried polymer samples placed in the opened capsules and then the heat flow-temperature curves in two heating-cooling cycles were measured. The sapphire crystal was used as standard. The heat capacity-temperature DSC thermograms calculated from these data are shown in Figures 1, 2.

DSC thermograms of amorphous PAA (Figure 1, a) contain the intense endothermic peak of water evaporation and one glass transition. On the DSC thermograms of crystalline PEO (Figure 1 b, c) only melting peaks are revealed. The temperature (T_m) and enthalpy (ΔH_m) of the melting transition increases with growth M_{PEO} but reduces in the 2-nd runs (Table 2).

DSC thermograms of the triblock copolymers with relatively shot PEO blocks ($M_{vPEO}\leq4\cdot10^4$) do not contain any melting transitions for PEO (Figure 1, e, f). A weak melting peak appears only in the thermogram of PAA-b-PEO-b-PAA3 with the longest PEO block (Figure 1, g). The ΔH_m value for this melting transition (Table 2) is much lesser than that one for melting process in individual PEO2. This fact testifies to much lesser crystallinity of the PEO block in the copolymer comparing with individual PEO. Moreover, the T_m value found for the given copolymer essentially reduces in comparison with analogous value for PEO2 (Table 2), that is the phenomenon of the melting point depression takes place. It is known that above-mentioned effects namely: the melting point depression and decreasing of the crystallinity degree of a polymer in a polymer mixture are conditioned by strong interaction of polymer components in the amorphous regions of a bulk structure.[5]

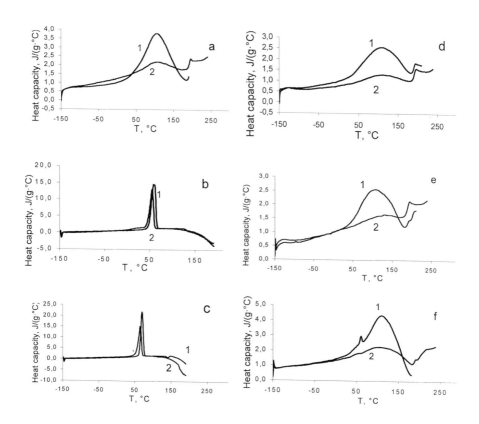

Figure 1. DSC thermograms of PAA –a, PEO1 –b, PEO3 –c, PAA-*b*-PEO-*b*-PAA1 – d, PAA-*b*-PEO-*b*-PAA2 –e, PAA-*b*-PEO-*b*-PAA3 –f. 1-st run (1) and 2-nd run (2).

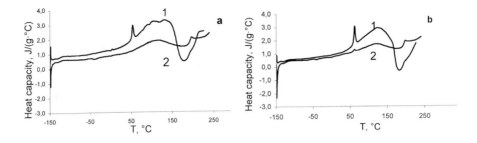

Figure 2. DSC thermograms of polymer blend of PAA+PEO1 –a and polymer blend PAA+PEO2 – b. 1-st run (1) and 2-nd run (2).

Table 2. Parameters of structural transitions in PAA, PEO and PAA-*b*-PEO-*b*-PAA.

Polymer	Cycle	T_g,	ΔT_g,	ΔC_p [1],	T_m,	ΔT_m,	ΔH_m,	X_c [2],
	(run)	°C	°C	J/g·°C	°C	°C	J/g	%
PAA	2-nd	190.9	8.0	0.55	-	-	-	-
PEO1	1-st	-	-	-	58.0	78.0	157.6	80
	2-nd	-	-	-	55.0	105.1	141.1	72
PEO2	1-st	-	-	-	70.0	101.0	172.3	88
	2-nd	-	-	-	65.0	100.0	156.3	79
PAA-*b*-PEO-*b*-PAA1	2-nd	190.8	9.0	0.55	-	-	-	-
PAA-*b*-PEO-*b*-PAA2	1-st	187.2	9.3	0.24	-	-	-	-
		202.0	8.5	0.34				
	2-nd	188.0	12.3	0.57	-	-	-	-
PAA-*b*-PEO-*b*-PAA3	1-st	-	-	-	60.9	24.0	73,2	37
	2-nd	187.0	6.6	0.33	56.1	30.0	22.5	11
		204.6	20.8	0.72				

[1] The heat capacity jump
[2] The degree of crystallinity. $X_c = \Delta H_m / \Delta H^\circ_m$, where ΔH°_m is the melting enthalpy of 100% crystalline polymer (196.8 J/g).[6]

It allows to conclude that the PAA and PEO blocks strongly interact with each other in the triblock copolymers. Comparative analysis of the parameters T_g and ΔT_g and ΔC_p, which characterize thermal transitions in the amorphous regions of the triblock copolymers and individual PAA, confirms such conclusion.

Really, DSC thermograms for PAA-*b*-PEO-*b*-PAA2 and 3 samples (Figure 1, f, g) demonstrate by contrast with thermogram of the 1-st copolymer sample (Figure 1, e) two glass transitions. The T_g value of the first of them is lower than that of individual PAA (Table 2) but the T_g of the second glass transition is higher. Thus, a microphase separation in the amorphous regions of the structure of both considered copolymers occurs. Analogous situation was observed in the bulk structure of the PAA to poly(vinyl alcohol) graft copolymers (forming IntraPC) since some critical values of the number and molecular weight of the grafts.[1] Such sort of microphase separation was attributed to existence in the amorphous areas of the graft copolymer structure side by side with the regions of full compatibility of polymer components also domains formed by PAA segments only which do not contact with the main chain. Basing on these data one can suppose that appearance of the microphase separation in the triblock copolymer amorphous regions is connected not with lengthening of the PEO block (that can only intensify the cooperative interactions between different blocks), but mainly with corresponding growth of the PAA block length (Table 1). Similar but more weak effects observe for two bulk polymer mixtures having the same polymer rations as in corresponding triblock copolymers (Figure 2, Table 3). These mixtures were obtained by freeze drying of the mixture of aqueous solutions.

Table 3. Parameters of structural transitions in PAA + PEO blends.

Polymer	Cycle (run)	T_g, °C	ΔT_g, °C	ΔC_p, J/g·°C	T_m, °C	ΔT_m, °C	ΔH_m, J/g	X_c, %
PAA+PEO1	1-st	189.5	9.2	0.58	53.4	25.0	160.1	81
		208.5	11.8	1.00				
	2-nd	193.0	9.0	0.62	-	-	-	-
PAA+PEO2	1-st	191.9	9.1	0.68	62.3	29.0	146.9	75
		-	-	-				
	2-nd	193.3	8.0	0.61	60.9	19	12.9	7

Really, the depression of T_m values of PEO in both mixtures and lowering of X_c (in the PAA+PEO2 mixture) comparing with such parameter in individual PEO are displayed in the 1-st heating cycle (1-st run). Additionally, two glass transitions are revealed in the 1-st run for both polymer mixtures (Figure 2 a, b; curves 1). Compatibility of the polymers essentially grows after heating of their mixtures higher than T_g. Such conclusion is confirmed by practically full absence of the PEO melting peaks and appearance only one glass transition in the 2-nd runs for both polymer mixtures (Figure 2 a, b; curves 2; Table 2).

Interaction between PEO and PAA blocks in TBC was studied by IR spectroscopy. IR spectra of thin films of individual polymers and two TBC samples, were recorded with the use of FTIR Micolett Nissan – 450 (USA) spectrometer. The examples of spectra are shown in the Figure 3.

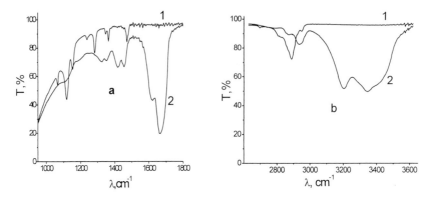

Figure 3. IR spectra of PEO2 –*1* and PAA-*b*-PEO-*b*-PAA2 –*2* films in the most important regions (a, b). T=293 K.

Formation of the H-bond system between polymer blocks is proved first of all by the low-frequency displacement (in 23 cm^{-1}) of the v_{C-0-C} vibration band of PEO in TBC comparing with the same band in the spectrum of individual PEO (v_{C-0-C} 1113 cm^{-1}). In order to establish the changes in the H-bond system between amide groups of PAA in TBCs (in comparison with individual PAA) under influence of the central PEO block the computer dividing of the overlapping amide I bands using the spline-functional method was carried out.[7] One examples is shown in Figure 4.

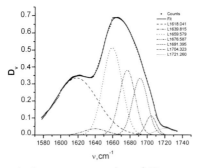

Figure 4. Computer processing of IR spectra of PAA-*b*-PEO-*b*-PAA2 in the amide I and amide II region. Experimental (···) and calculated (—) vibration band contours. The separate amide I bands are shown also.

Using only two overlapping bands in the amide I and amide II region it can not be correctly process these areas of spectra. Therefore, all bands of PAA which were well known and described earlier were introduced into a computer.[8] Then the integral absorption coefficients of separate amide I bands were used to calculate their contributions (α_i) in the common carbonyl absorption (Table 4).

Table 4. Contributions of the apparent integral absorption coefficients of separate amide I bands in common $v_{C=O}$ absorption.

Polymer	$\dfrac{\alpha}{\%}$				β [1]$=$ $B_{1676}/$ B_{1691}
	$v{\sim}1660$ cm^{-1}	$v{\sim}1676$ cm^{-1}	$v{\sim}1691$ cm^{-1}	$v{\sim}1704$ cm^{-1}	
PAA	47.6	23.7	24.8	4.0	0.95
PAA-*b*-PEO-*b*-PAA2	45.9	29.4	19.1	5.7	1.54
PAA-*b*-PEO-*b*-PAA3	45.3	26.7	22.9	5.1	1.17

[1]The effective length of *trans*-multimers of amide groups

Note, that the band near 1660 cm^{-1} is attributed to *cis-trans*-multimers of amide groups.[8] Two bands at 1676 cm^{-1} and 1691 cm^{-1} belong to terminal and inside (H-bond) amide groups of *trans*-multimers. Finally, the band at 1704 cm^{-1} points on existence of free amide groups.

Basing on the data of Table 4, it can be conclude that there are relatively lesser of *cis-trans* and also *trans*-structures of H-bound amide growps in the TBC samples under study comparing with individual PAA. At the same time the effective length of *trans*-multimers of amide groups in the TBC samples, in comparison with PAA, grows. Increasing of rigidity of PAA chains in TBCs can be explained by a participation of *trans*-multimers of the PAA amide groups in the formation of H-bond system between PEO and PAA blocks, as in the case of intermolecular complex formation between PAA and poly(vinil alcohol):[9]

Conclusion

Thus, triblock copolymers based on PEO and PAA are intramolecular polycomplexes, stabilized by the system of H-bond between PEO and PAA blocks. Intramolecular complex formation is reflected in the practically full absence of PEO crystalline domains in the block copolymer structure and also in the specific microphase separation in amorphous domains.

[1] T.B. Zheltonozhskaya, N.E. Zagdanskaya, O.V. Demchenko, L.N. Momot, N.M. Permyakova, V.G. Syromyatnikov, L.R. Kunitskaya, *Usp. Khim.* **2004**, *73*, 877, [*Russ. Chem. Rev.* **2004**, *73*,…(in press)]

[2] T. Zheltonozhskaya, O. Demchenko, I. Racovich, J.-M. Guenet, V. Syromyatnikov, *Macromol. Symp.* **2003**, *203*, 173.

[3] O. Demchenko, T. Zheltonozhskaya, J.-M. Guenet, S. Filipchenko, V. Syromyatnikov, *Macromol. Symp.* **2003**, *203*, 183.

[4] N.P. Melnik, L.N. Momot, I.A. Uskov, *Dokl. Akad.Nauk Ukr. SSR*, **1987**, Ser. B, 6, 48.

[5] Wenwei Zhao, Li Yu, Xiaognang Zhong, Yuefand Zhang, Jiazhen Sun, *J. Macromol. Sci. - Phys.* **1995**, *B34*, 231.

[6] V.A. Bershtein, V.M. Yegorov, *Differentsyalnaya Skaniruyushchaya Kalorymetriya v Phyziko-Khimii Polymerov*, Khimiya, Leningrad. **1990**, s.256.

[7] G.Nurberger, *Approximation by Spline Function*, Springer, Verlag. **1989**, p.243.

[8] N. Permyakova, T. Zheltonozhskaya, O. Demchenko, L. Momot, S. Filipchenko, N. Zagdanskaya, V. Syromyatnikov, *Polish J.Chem.* **2002**, *76*, 1347.

[9] T.B. Zheltonozhskaya, *Doctorel Thesis in Chemical Science*, Kyiv, National University, Kyiv, **2003**.

Macromol. Symp. **2005**, *222*, 143-148   143

Identification of Gel-Like Behaviour in Side-Chain Liquid Crystal Polymer Melts

Hakima Mendil,* Laurence Noirez

Laboratoire Léon Brillouin (CEA-CNRS), Ce-Saclay, 91191 Gif-sur-Yvette Cédex, France

Summary: We show through extensive rheological studies that Side-Chain Liquid Crystalline Polymer melts reveal an unexpected and surprising strong elasticity instead of a classical flow behaviour. Neutron Scattering experiments demonstrate that this elastic plateau cannot be correlated to the long range order of the nematic phase.

Keywords: birefringence; gel behaviour; liquid-crystalline polymer; neutron scattering; non-linear viscoelasticity; shear induced transitions

Introduction

Thermotropic Liquid Crystal Polymers (LCPs) are attractive materials both to academic research and to industrial applications [1]. The present letter is focused on side-chain LCPs. In such a case, the liquid crystal molecules are grafted, for example as an ester group, on the side of an ordinary polymer chain.

Here we report on a comparative analysis of the mesophase structure carried out by neutron scattering with an extensive non-linear rheological investigation. The polymer chosen is a low polydisperse liquid crystalline polyacrylate (PACN) of chemical formula:

Mw =91 000 g.mol^{-1}, Ip=1.1

It displays the following mesophase sequence: *I(isotropic) - 119°C – N(nematic) - 30°C - Glassy state* (temperatures determined by Differential Scanning Calorimetry).

These rheological studies reveal a spectacular and unexpected cohesion within the melt. This striking behaviour demonstrates that LCP melt behaves as a gel. The shear induced nematic-isotropic phase transition identified in 2001[2] on Side-Chain-LCP melts can be explained as a consequence of the deformation of an elastic structure. Neutron Scattering measurements demonstrate that the gel behaviour cannot be due to the long range order of the mesophases.

        DOI: 10.1002/masy.200550417

Rheology

Viscoelastic measurements were carried out with an ARES rheometer equipped with an air-pulsed oven and used in dynamic frequency sweep mode. This thermal environment ensures temperature control within 0.1°C. The samples were placed between a cone-plate fixture (20mm diameter and angle 2.25°) and thermalised during several hours before starting the measurements.

Figure 1 displays the typical frequency dependence of the elastic modulus $G'(\omega)$ and $G''(\omega)$ at strain amplitudes from $\gamma=1$ up to 100%. The rheological spectrum reveals a completely new behaviour: the LCP exhibits a pseudo-solid-state behaviour instead of a classical flow regime.

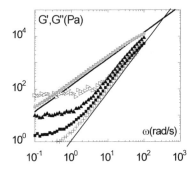

Figure 1. Frequency dependence of the viscoelastic moduli ($G'(\omega)$ and $G''(\omega)$) at T-$T_{NI}=\Delta T=+1$°C (cone-plate geometry (20mm diameter and angle 2.25°)) at different strain amplitude γ :($\triangleright$)1% ($\blacktriangle$)5% ($\blacksquare$)50% (+)100%. The straight lines display the extrapolation to ω and ω^2 scaling.

At low strain ($\gamma=1$%) the magnitude of the elastic plateau G_p' reaches 10^2Pa. The larger the strain amplitude, the lower is G_p'. It defines a non-linear frequency regime. In such conditions, the time-temperature superposition fails. The extrapolation of $G'_p(\gamma)$ to zero strain rate gives an estimation of the non-disturbed structure; $G'_0(\gamma=0)$ is about $10^{5\pm1}$Pa. This elasticity persists below (-14°C), through and far above (at least +15°C) the Isotropic-Nematic transition temperature T_{NI}. The high magnitude of the elastic plateau, its persistence through and far from the transition indicate that this gel behaviour should not be coupled with pretransitional dynamics [3].

Except true rubbers, systems known to display low-frequency elastic plateaus are heterogeneous. This is the case of gelation processes, for instance in associative polymer systems or in copolymer solutions (G_p '#10^{4}Pa) [4]; viscoelastic emulsions (G_p '#10^2-10^{3}Pa) [5]. In these examples, the low frequency elastic plateau comes out from an heterogeneous structure or a transition ordering. Such arrangement is starting at local scale and propagating to the whole sample. Our system is fundamentally different since it is chemically homogeneous. We observe a novel type of cohesion not due to chemical heterogeneities or a cross-linking. We checked of course that the molar weight unchanges before and after the rheological investigations.

At high frequency and high strain, the classical polymer melt behaviour is recovered: $G''(\omega)$ and $G'(\omega)$ fit with ω and ω^2 scaling respectively. The non-linearity shows that the gel-like behaviour becomes viscoelastic under increasing strain; we propose to describe the elastic modulus behaviour by two terms: $G'(\omega, \gamma) = G'_p(\gamma) + \eta.\tau_{term}.\omega^2/(1 + \omega^2.\tau^2)$ where τ_{term} is the viscoelastic terminal time. The first term accounts for the non-linear elasticity, whereas the second contains the conventional viscoelastic contribution. τ_{term} is deduced from the intersection of the two straight lines; $\tau_{term}(\Delta T=+1°C) \cong 4.10^{-3}$s where ΔT is the temperature interval T-T$_{NI}$. The insensitivity of $G''(\omega)$ down to 10rad/s indicates that the material keeps a viscous component coherent with a gel behaviour.

This observation leads to the conclusion that this cohesion is apparently not reminiscent from the mesophases. Is it connected to a supramolecular structure unrevealed till yet? Neutron scattering measurements are carried out to characterize the structural organization within the liquid crystal melt over a wide range of temperature below and above the Isotropic-Nematic transition.

Neutron Scattering

Small Angle Neutron Scattering (SANS) was performed on the PAXY spectrometer of the Laboratoire Léon Brillouin. A wavelength of 4Å and a sample-multidetector distance of 1,5m were chosen. A magnetic field of 1.4T was used to ensure the alignment of the sample.

The identification of structural arrangements and of their temperature dependence can be characterized by diffraction. We will show that the evolution of the structure of the mesophases is independent of the non-linear elasticity exhibited by the sample.

Figure 2 displays the 2D-SANS pattern obtained at ΔT=-20°C.

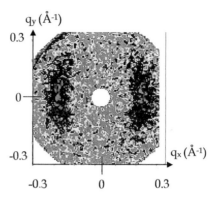

Figure 2. 2-D neutron scattering pattern (sample-detector distance: 1.5m, wavelength: 4Å covering a scattering range from 0.03Å^{-1} up to 0.3Å^{-1} displayed at $\Delta T=-20°C$ (nematic phase) by a monodomain sample oriented with the director parallel to the Ox axis (PAXY spectrometer at the LLB).

In the reciprocal space, we observe a diffracted intensity at a distance $d=2\pi/q=28.6\text{Å}$. It is due to S_{Ad} smectic fluctuations within the nematic phase. Since the molecular length of the mesogen is about 17Å, it corresponds to a partial overlap.

On Figure 3, the amplitude and the scattering vector of the S_{Ad} smectic fluctuation intensity are reported as a function of the temperature. As already reported [6] in systems displaying monolayer smectic phase (S_{A1}), the scattering vector amplitude slightly increases with the temperature. It has been interpreted in terms of restriction of the motion of the side-chain. The intensity of the smectic fluctuations decreases as the temperature increases: the local mesogen arrangement leading to a local smectic order weakens. At the Isotropic-Nematic temperature transition, this intensity drops close to zero and the associated scattering vector diverges as expected.

The scattering pattern above the I-N transition looks similar to every ordinary isotropic phase of liquid crystals and does not contain any information on structural arrangement within this scattering observation window. In contrary, as previously discussed, the elastic plateau is insensitive to the I-N transition. Both antagonist behaviours demonstrate that the elastic plateau cannot be correlated to the long range nematic order or other supra molecular arrangement.

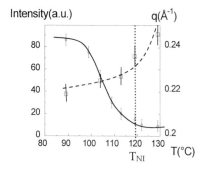

Figure 3. Diffracted intensity and scattering vector of the S_{Ad} smectic fluctuations as a function of the temperature. One can notice the abrupt decrease of the intensity and increase of q at the Isotropic-Nematic temperature transition. The straight, dotted and dashed lines are guides for the eyes.

Conclusions

Instead of a conventional flow behaviour, we have observed a strong cohesion within the liquid crystal polymer melt. These observations imply that extra long length and time scales exist. The confrontation with the structural study by neutron diffraction shows that this huge elasticity does not originate from mesomorphic properties. This is a melt property which should be observable in other viscoelastic materials [7] .

This study extends at a macroscopic scale and in the non-linear regime, the very interesting pioneering work evidencing a gel-like behaviour in polymers on low thickness samples [8]. Far above phase and glass transitions, these long range interactions may explain spectacular non-linear phenomena as the shear induced Nematic-Isotropic phase transition in LCPs [2]. This transition is revealed by the abrupt emergence of birefringence above a critical shear rate $\dot{\gamma} > \dot{\gamma}^*$ (Fig.4). Figure 4 illustrates the evolution of the birefringence at $\Delta T=+1°C$ in steady shear conditions. To explain such a phenomenon, a direct coupling with the life time of the pretransitional fluctuations is not relevant [3,9]. A coupling with the viscoelastic terminal time τ_{term}, is also not satisfying since $1/\dot{\gamma}^*$ is significantly larger than $\tau_{term}(\Delta T=+1°C) \cong 4.10^{-3}s$. A classical approach is not sufficient to elucidate neither the huge melt cohesion nor the spectacular shear induced transition. Any theoretical attempt should indeed account for the following time scale cartography: $1/\dot{\gamma}^* >> \tau_{term} > \tau_{fluc}$. Although the liquid crystal polymer melt does not correspond to the definition of an entangled melt (low degree of polymerisation and absence of rubbery

148

plateau at high frequency), theories developed for elastomers [10] and for branched polymers [11] could be very challenging approaches for the understanding of these non-conventional behaviours.

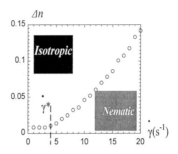

Figure 4. At $T=T_{NI} +1°C$ and $\dot{\gamma} > \dot{\gamma}*$, emergence of the shear induced nematic phase within the isotropic phase (observation between cross-polarisers, plane: velocity, neutral axis, gap thickness $e=100\mu m$).

Acknowledgements

The authors are very grateful to P. Baroni for his assistance during experiments, and for the adaptation of the optical shear set-up to our experimental conditions.

[1] "Handbook of Liquid Crystals", Wiley (2001).
[2] C.Pujolle-Robic, L.Noirez, Nature **409** (2001) 167.
[3] S. Hess, Z. Naturforsch. **31**a (1976)1507; P.D. Olmsted, P. Goldbard, Phys. Rev. A**41** (1990) 4578; ibid A**46** (1992) 4966.
[4] Y. Aoki, Macromol. **20** (1987) 2208 ; G. Tae et al, Macromol. **35** (2002) 4448, H. Watanabe et al Macromol. **34** (2001) 6742.
[5] M. Bousmina, Rheol.Acta **38** (1999) 251.
[6] L.Noirez, P.Davidson, W. Schwarz, G.Pépy, Liquid Crystals, **16** (1994) 1081.
[7] H.Mendil, P.Baroni, L.Noirez submitted to PRL; L. Noirez, H. Mendil, I. Grillot submitted to Europhys. Lett.
[8] J.L. Gallani, L. Hilliou, P. Martinoty, P. Keller, Phys. Rev. Lett. **72** (1994) 2109; P. Martinoty, L. Hilliou, M.Mauzac, L. Benguigui, D. Collin, Macromol. **32** (1999) 1746; D. Collin, P. Martinoty, Physica A **320** (2002) 235.
[9] V. Reys, Y. Dormoy, J.L. Gallani, P. Martinoty, P. Lebarny, J.C. Dubois, Phys. Rev. Lett. **61** (1988) 2340.
[10] M. Warner, K.P. Gelling, J. Chem. Phys. **88** (1988) 4008, A. Halperin, J. Chem. Phys. **85** (1988) 1081; R. Sigel, W. Stille, G. Strobl, R. Lehnert, Macromol. **26** (1993) 4226.
[11] T.C.B. McLeish, S.T. Milner, Advances in Polymer Science, **143** (1999)

Macromol. Symp. **2005**, *222*, 149-155

149

Structure of Hydrogen and Hydrophobically Bonded Amphiphilic Copolymer with Poly(methacrylic acid) Complexes as Revealed by Small Angle Neutron Scattering

Mehdi Zeghal,[*1] *Loïc Auvray,*[2] *Mondher Jebbari,*[3] *Abdelhafidh Gharbi*[3]

[1] Laboratoire de physique des solides, Université Paris XI, 91400 Orsay, France

[2] Laboratoire des Matériaux Polymères aux Interfaces, Université d'Evry Val d'Essonne, 91025 Evry, France

[3] Laboratoire des Cristaux Liquides et des Polymères, Département de Physique, Faculté des Sciences de Tunis, 1060 le Belvédère, Tunisie

Summary: We study by SANS the structure of intermolecular complexes formed through hydrogen bonding and hydrophobic interactions between poly(methacrylic acid) (PMA) and a neutral copolymer surfactant (PEO-PPO-PEO). The contrast variation method enables us to probe the structure factor of each polymer in the complex and their cross structure factor. The number of copolymer chains, which results from the cooperative action of hydrogen bonding and hydrophobic interactions increases as the charge of the polyacid decreases. The aggregation preserves the micellar core-corona organization of the copolymer and shrinks the polyacid chains which adopt a similar compact structure. Finally, the structure of the aggregates is compared to that of PEO-PMA homopolymer complex observed by SANS.

Keywords: amphiphiles; SANS; surfactants

Introduction

Intermolecular forces that drive self association in biological or soft condensed matter systems in water are mainly the electrostatic, the hydrogen bonding and the so-called hydrophobic interaction. In most of cases, these effects operate together in a cooperative way [1,2,3,4], and it is difficult to estimate the contribution and the effect of each type of interaction on the structure of the complex. For that reason, we decided to investigate the structure of a polymer complex stabilized simultaneously via hydrogen binding and hydrophobic interactions. Our choice was fixed on a complex formed by an amphiphilic copolymer (PEO-PPO-PEO) [2] and poly(methacrylic acid) that can form hydrogen bonds between the hydroxyl groups of the polyacid and the oxygen of the poly(ethylene oxide) (PEO) or the poly(propylene oxide) (PPO) [1-3-4]. Furthermore, the hydrophobicity of PPO

© 2005 WILEY-VCH Verlag GmbH & KGaA, Weinheim   DOI: 10.1002/masy.200550418

and non ionized parts of poly(methacrylic acid) that depend essentially on temperature can also promote aggregation. Beside, the repulsive electrostatic interactions and the entropy of the counterions act on the opposite way and resist aggregation. In order to sort out these relative contributions to the final structure of the complex, we decided to study the structure of the complex using Small Angle Neutron Scattering (SANS) and more specifically, the contrast variation method which allows to probe the structure of each polymer and their mutual interactions[3]. We focus our attention on the effect of the polyacid neutralization degree α on the structure of the complex.

Materials and methods

Fully deuterated poly(methacrylic acid)(PMA-D, M_w=370000 g/mol) was synthesized by radical polymerization of deuterated methacrylic acid. The degree of neutralization of the polyacid (α=[COO$^-$]/([COO$^-$]+ [COOH])) was adjusted by adding the required quantity of NaOH. The copolymer, known by its trade name as Pluronic$^®$ (P105, EO_{37}-PO_{56}- EO_{37}) was a gift from BASF (France). It was used without further purification.

The scattering experiments were performed at Laboratoire Léon Brillouin on the spectrometer PACE. The range of scattering vector was 10^{-2} Å^{-1}< q < 0.11 Å^{-1}. The measured normalized scattering intensity I(q) (in cm^{-1}) is related to the partial structure factors of the sample constituants by the equation.

$$I(q)= (n_a-n_w)^2 S_{aa}(q)+ (n_a-n_w) (n_b-n_w)S_{ab}(q)+ (n_b-n_w)^2 S_{bb}(q) \quad (1)$$

n_a, n_b and n_w are respectively the scattering length densities of PMA(D), the copolymer and of the solvent. The partial structure factors S_{ij} are defined as usual by $S_{ij}(q)= \int d^3r \langle \delta\Phi_i(0)\delta\Phi_j(r)\rangle e^{iqr}$, where Φ_i(r) is the local volume of constituent i and $\delta\Phi_i$(r) the local deviation from the average value in the sample volume. The partial structure factors S_{ii}(q) are related to the structure of each polymer in the complex, whereas the cross structure factor S_{ab}(q) gives information on the interaction between the constituents of the complex. The contrast variation experiments were performed by measuring the scattering intensity of the solutions prepared with three proportions of heavy water and the three partial structure factors are obtained by solving eq. 1 for these three compositions of the solvent [4]. SANS experiments were performed at 30°C; the copolymer concentration was 2.5% (w/v) and the ratio [PEO]/[PMA]=2.

Results and discussion

At 30°C, well above the critical micellization temperature (about 20°C), the copolymer form micelles with a dry core of PPO surrounded by a hydrated coronna of PEO [2]. The intensity scattered by the micelles is fitted using a hard sphere interaction structure factor and a core-corona form factor [4]. The fitting procedure is based on four fitting parameters: N_a: the micelle association number, R_1: the core radius, R_2: the corona radius and R_{hs}: the hard sphere interaction distance. The result of the scattered intensity fit is displayed on figure 1, giving R_{hs}= 100 Å, R_1= 42.5 Å and R_2= 59.1 Å. The intermicellar interaction radius is 60% larger than the micelle radius, indicating that the micelles are well separated from each other [4].

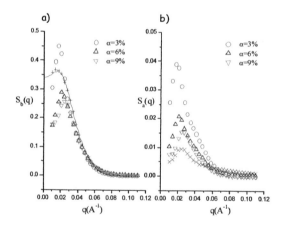

Figure 1. a) Structure factor of the copolymer in the complex for three values of α. The solid line represents the best fit of the structure factor of the pure copolymer in water (+). b) Structure factor of the PMA(D) in the complex for three values of α and alone in water for α = 6% (□).

The study of the increasing of the scattered intensity remains the simplest way of characterizing aggregation as the association proceeds, and usually, the larger the intensity, the larger the aggregates. This is indeed what we observe in figure 1 where we compare the structure factors of PMA and the copolymer chains in the complex with their counterparts in the pure equivalent solution. Precisely, we observe that the intensity scattered by each polymer in the complex increases as the degree of neutralization α decreases. In other words, the aggregation number of each polymer in the complex increases with the density of non ionized (complexable) hydroxyl groups along the polyacid chains. This is confirmed by the positive cross structure factors $S_{ab}(q)$ (figure 3) which indicates that the interaction between the copolymer and the PMA is attractive. In addition, this attractive interaction increases with the density of available carboxyl groups as reflected by the evolution of $S_{ab}(q)$ with α.

The number of PMA chains in the complex decreases with the density of complexable monomers along the polyacid chains and the comparison of the scattered intensities allows to give an estimation of this number that varies from 2 chains (α=9%) to 4 chains (α=3%) per aggregate.

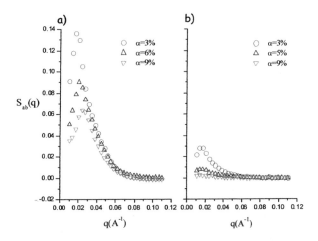

Figure 2. a) Cross structure factor of the copolymer-PMA(D) complex in water for three values of α. b) Cross structure factor of PEO-PMA(D) in the complex in water for three values of α.

The number of copolymer chains per aggregate also decreases with the density of complexable groups on PMA chains. More precisely the scattered intensities indicate that this number is divided by a factor of about 2 in the range of variation of α. It is interesting to notice that at high neutralization, the number of copolymer chains is surprisingly lower in the complex than in the micelles. An additional interesting feature is that the copolymer correlation peak is not only shifted in comparison with the micelle peak position, but its shape changes from a smooth correlation peak reflecting the hard sphere interaction potential to a sharp electrostatic peak due to long range Coulombic interactions between aggregates. Furthermore, for a given α value, the position of the electrostatic peak observed on the three partial structure factors is that of the free PMA chains at the same concentration, indicating that the aggregates repel each other via electrostatic interactions as free PMA chains [4].

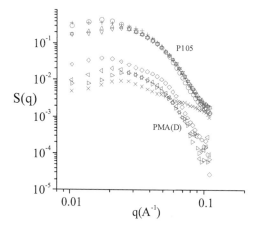

Figure 3. Structure factor of the copolymer in the complex for three values of α ($\circ$:α=9%, $\lceil$:α=6%, $|$:α=3%) and alone in water (+). Structure factor of the PMA(D) in the complex for three values of α ($\cong$:α=9%, 0:α=6%, $\langle$:α=3%) and alone in water for α = 6% ($\square$).

The shape of the polymer partial structure factors at large scattering vector is related to the local structure of the chains within the complex. Beyond the correlation peaks, the intensity scattered by the copolymer in the micelles or in the complex is the same whatever the value of α (Figure 3). It appears that the complexation does not change the organization of the copolymer compared with the micelles whatever the strength of the interaction (i.e. the value of α). The preservation of the core-corona structure of the copolymer in the presence of PMA is confirmed by dynamical information obtained by NMR [4,5], showing that the PEO block dynamics is not affected by complexation and forms the hydrated corona, whereas the PPO block dynamics is strongly hindered by the association with PMA, forming the dry core of the aggregate. One can notice that, unlike the copolymer, the structure of PMA chains in the presence of copolymer and alone in water are very different (Figure 3). Precisely, in the high q regime, the intensity scattered by PMA decreases more rapidly for the complex than for free chains, showing that the association process modifies the local structure of the polyacid up to the molecular scale by shrinking the chain. The scattered intensity decay does not depend on α and is very similar to that of copolymer chains, indicating that the core-corona picture is also valid to describe polyacid chains organization within the complex [4]. In that scheme, most of non-dissociated carboxyl group sequences are complexed with the PPO, forming the dense hydrophobic core of the complex. This core is surrounded by a hydrated corona formed with free ionized polyacid sequences and PEO blocks.

It is interesting to compare the structure of the copolymer-PMA complex with that of PEO-PMA homopolymer complex obtained with the same contrast variation method [3]. Such a comparison allows to estimate the respective influence of the hydrophobic and the hydrogen bonding attractive interaction on the formation and the structure of the complex. The PEO-PMA complex was composed with the same polyacid and with a poly(ethylene oxide) (M_w=100000 g/mol) purchased from Aldrich [3]. In both cases, the aggregation number of each polymer and the size of the aggregates increase when α diminishes. These observations are confirmed by the increase of the cross structure factors as α decreases (Figure 2.), indicating that the attractive hydrogen bonding interaction increases with the density of complexable (non dissociated) carboxyl groups. The comparison between cross structure factors displayed in fig. 3 shows that, as the polyacid charge is higher than 9 %, there is no interaction between homopolymers ($S_{ab} \approx 0$) whereas the copolymer-PMA interaction is attractive ($S_{ab} > 0$). For the copolymer, the combined action of hydrogen bonding and hydrophobic effects allow to overcome the repulsive interactions. This is no more the case with the homopolymer for which the hydrophobic interactions are absent and the hydrogen bonding is too weak to overcompensate the repulsion between the polymers. Another important difference that deserves to be discussed between the two complexes is the evolution of the structure of each polymer in the aggregates when α decreases. Contrary to the copolymer-PMA complex, the structure of each polymer in the homopolymer complex depends on the degree of neutralization α [3]. For α=9%, no complexation occurs and the structure factors of the polymers is close to that of their counterparts alone in water at the same concentration. As α decreases, the structure factors of the free and complexed chains become progressively different, and for the lower values of α, the PEO and PMA chains appear to be zipped together, adopting the same compact structure [3]. The proportion of hydrophobic sequences formed with non dissociated PMA blocks and PEO increases as the neutralization of the polyacid decreases, leading to a progressive shrinking of the chains. Contrary to the complex formed with the copolymer, the local structure of the chains within the aggregates is very sensitive to the PMA neutralization.

Conclusion

The contrast variation method is a powerful tool to probe the structure of the intermolecular complex formed with a polyacid and a copolymer. Above the cmt, the aggregation number of each polymer increases when the density of ionized PMA groups decreases. The aggregation number of the copolymer decreases below that of pure micelles if the polyacid charge exceeds 3-4%. It appears that the complexation preserves the core-corona organization of the copolymer and shrinks the PMA chains that adopt a similar core-corona structure. Contrary to what happens with PEO-PMA homopolymer complexes, the structure of both constituents is not sensitive to the polyacid charge. Whatever the neutralization, the non dissociated PMA blocks associate with PPO to form the dry core of the complex whereas the PEO and ionized blocks of PMA form the stabilizing hydrated corona. This polymeric system gives a particular demonstrative example of interaction between a surfactant and a polymer. Furthermore, it can help us to understand the association between biological polymers like DNA and Pluronic used in gene therapy [6]. Finally, at higher micellar concentration, clusters of interconnected micelles with long polyacid chains appear [7], leading to the formation of a novel class of pH and temperature sensitive gels.

[1] I. Iliopoulos, R.Audebert, J. Polym. Sci., **26**, 275 (1988).
[2] L. Yang, P. Alexandridis, D.C. Steyler, M.J. Kositza, J.F. Holzwarth, Langmuir, **16**, 8555 (2000)
[3] M. Zeghal, L. Auvray, Europhys. Lett., **45** (4),PP. 482-487 (1999).
[4] M. Zeghal, L. Auvray, Eur. Phys.j. E **14**, 259-268 (2004).
[5] F.Cau, S. Lacelle, Macromolecules, **29**, 170, (1996).
[6] B. Pitard, H. Pollard, O. Agbulut, O. Lambert, J.-T. Vilquin, Y. Cherel, J. Abadie, J.-L. Samuel, J.-L. Rigaud, S. Menoret, I. Anegon, D. Escande, Human Gene Therapy, **13**, 1767 (2002).
[7] M. Zeghal, M. Jebbari, L. Auvray, A. Gharbi, in preparation.

Macromol. Symp. **2005**, *222*, 157-162

Synthesis and Characterization of New Polymer Nanocontainers

Sandrine Poux, Wolfgang Meier*

Department of Chemistry, University of Basel, Klingelbergstrasse 80,CH-4056, Basel, Switzerland
Fax: +41 (0) 612673855; E-mail: wolfgang.meier@unibas.ch

Summary: New type of reactive, water-soluble and filled polymer nanocontainers that can be covalently attached to surfaces were synthesized. The encapsulation of a dye inside the nanocontainers allows their rapid detection. The model systems are based on crosslinked polystyrene (PS) and polymethylmethacrylate (PMMA). The synthesis of nanocontainers by two-step emulsion polymerization and characterization by NMR, IR, Dynamic Light Scattering, TEM and Confocal Laser Scanning Microscopy (CLSM) is presented.

Keywords: core-shell polymer; hollow sphere; nanoparticle

Introduction

In recent years, considerable progress has been made in the development of synthetic methods allowing the preparation of materials with precise control over size and morphology at the nanometer level. The principal example is the formulation of hollow nanoparticles. These hollow particles are particularly interesting for applications as confined reaction vessels, drug carriers or protective shells for dyes used as labels for enzymes or catalyst.[1-2] Similar nanometer-sized containers, e.g., micelles or vesicles are used by nature in biological system. However, their limited mechanical stability prevents many possible applications (e.g., in drug delivery).[3] Mechanically stable polymer nanocapsules can be prepared by using multiple techniques (see, e.g. [4-9]). Among others, water soluble and filled polymer nanocontainers can be synthesized by two-step emulsion polymerization via core-shell latexes. Here, our final objective was to develop a new technology that allows a specific and selective immobilization of labelled nanocontainers at surfaces. We first synthesized a new type of reactive, water-soluble and filled polymer nanocontainers that can be further covalently attached to surfaces. The encapsulation of a fluorescent dye (carboxyfluorescein) inside the nanocontainers allows quick detection of the particles. The core-shell particles presented in this report contain a liquid polydimethylsiloxane (PDMS) core and a shell of poly(styrene-methylmethacrylate-

DOI: 10.1002/masy.200550419

divinylbenzene), Poly(St-MMA-DVB). This system allows convenient separation and purification of the shell forming polymer.[10] Furthermore, the preparation method is suited for scaling up the production of different container systems and allows the optimization of the particles by variation of particle size, shell thickness and crosslinking density.

Synthesis of polymer nanocontainers

As a model system, we used crosslinked PS and PMMA nanocontainers by using DVB as crosslinking agent. The particles were synthesized by two-step emulsion polymerization via core-shell particles (Figure 1).[11] In the last step, the PDMS core can be removed by ultrafiltration in toluene.

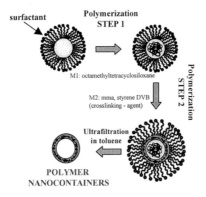

Figure 1. Schematic representation of the polymer nanocontainer preparation via core-shell latexes.

In the first step, the PDMS core is prepared by ring opening cationic polymerization of octamethyltetracyclosiloxane (M1) in water with dodecylbenzenesulfonic acid (DBSA) as both catalyst and surfactant. The synthesis of PDMS via the ring-opening polymerization of cyclic siloxane monomers has been described previously.[12-13] In general, this procedure yields polymers of low molecular weight. Concerning the alternative preparation route for this polymer, by emulsion polymerization, Bey *et al.* obtained PDMS latexes using DBSA as both surfactant and catalyst; the reported particle sizes varied between 50 and 500 nm with 1.5-4 g of DBSA for 100 g of siloxane monomer.[14-15] For that reason, the polymerization of PDMS was conducted using 20 g of water, 5 g of M1 and an appropriate amount of DBSA (between 0.5-5.0 wt% with respect to M1). Distilled water and the surfactant were added into a three-neck flask. The reaction mixture was magnetically stirred at ca. 400 rpm at room temperature and oxygen was removed by bubbling argon for 30 minutes. Next, the monomer was added dropwise. After the polymerization proceeded, the reaction was quenched with 1N NaOH.

In the second step, the synthesis of the crosslinked shell was done via the seeded polymerization of St-MMA-DVB on the seed PDMS particles. Potassium persulfate (KPS) was used as initiator, without additional surfactant. A selected amount of PDMS latex (10 mL), water (10 mL) and KPS (2.5 wt% relative to St-MMA-DVB monomers) were first loaded into the flask, from which oxygen was removed by bubbling argon for 30 minutes. When the temperature reached 80°C, the monomer mixture was added dropwise at a low rate (ca. 10 ml/h) and the polymerization was proceeding for 4 hours. Then, in the third step, the PDMS core was removed by dissolving the core-shell particles in toluene and subsequent ultrafiltration (Millipore cellulose 100000). Finally, concerning the encapsulation of carboxyfluorescein, two strategies were used. In the first approach, the nanocontainers were dissolved in a good solvent (e.g dioxane), where they swell and become permeable so that the dye molecule can diffuse into their interior from the solution. Then, a subsequent change of solvent conditions (e.g addition of water) will contract the polymer shells and decreases their permeability, while the dye molecules will be entrapped in the container interiors. By the second, more convenient strategy, carboxyfluorescein is incorporated directly during the preparation of core-shell particles. This way, the dye has to be dissolved within the liquid PDMS core. The main advantage of this approach is that the particles can be used directly without the last purification step, where the PDMS core is removed.

Characterization of polymer nanocontainers

The resulting core-shell particles were first characterized by FTIR spectroscopy (Figure 2). The spectrum reveals strong adsorption at 2950 cm^{-1} and 1750 cm^{-1} that corresponds to aliphatic C-H and carbonyl C=O stretches, respectively. A medium band at 1260 cm^{-1} was due to the siloxane stretching. Two strong bands in the 1100-1050 cm^{-1} region and at 801 cm^{-1} originate from siloxane vibration (Si-O-Si) and from the silicone adsorption, respectively. Finally, the strong band at 699 cm^{-1} corresponds to the phenyl ring of PS. The IR spectroscopy data shows clearly the presence of PDMS core and that of PS-PMMA shell.

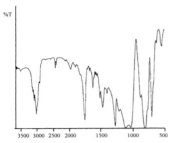

Figure 2. IR spectrum of the core-shell particles.

The size of the latex particles was determined by light scattering, which provided both the average diameter and the size polydispersity. Nanocontainers of different sizes were obtained by varying the monomer/surfactant (M1/DBSA) ratio in the first step of the synthesis (Table 1). A systematic variation of the particle size provides the information about the size dependence on the loading and detection of the nanocontainers containing a fluorescent dye. In fact, at the same concentration, larger particles will contain a higher number of dye molecules. However, concerning the immobilization and the distribution of labelled nanocontainers on surfaces, smaller particles allow a higher surface coverage and more homogeneous distribution.

Table 1. Particle size as a function of the dodecylbenzenesulfonic acid (DBSA) amount.

%weight of DBSA/M1	Dp^a (nm)	Polydispersity at $\theta=90$
0.5	470	0.24
1.5	336	0.08
2.5	246	0.22
3.5	110	0.19
5	86	0.13

[a] Particle diameters determined by dynamic light scattering after extrapolation to zero angle.

In the second step of the synthesis, we controlled the thickness of the container walls by varying the core-shell ratio in the reaction mixture (Table 2). In fact, a higher shell thickness leads to more stable and less permeable particles, but, at the same time, results in the more unfavourable encapsulation ratio.

Table 2. Shell thickness as a function of core-shell ratio in the second step of the emulsion polymerization.

Core/Shell molar ratio	%mol of crosslinking DVB/M2	Shell thickness[a] (nm)
1/1	10	20-25
1/0.5	10	5-7
1/0.25	10	-

[a] Shell thickness determined by dynamic light scattering.

The morphology of core-shell particles was studied by transmission electron microscopy (TEM). The PDMS/PS-PMMA-DVB core-shell particles were thinned with distilled water and sonicated. Uranyl acetate was used as the contrast enhancing stain. As can be seen in the TEM image (Figure 3 (a)), the synthesized particles were found spherical and very monodisperse in size. Their monodispersity is in good agreement with the dynamic light scattering data (Table 1). Furthermore, the image (Figure 3 (b)) shows clearly the formation of a core-shell particle, where it is possible to distinguish the core from the shell.

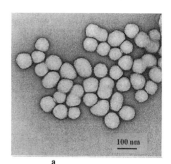 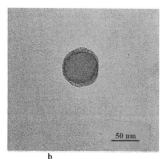

a b

Figure 3. TEM images of core-shell particles after the seeded polymerization of St-MMA-DVB on the seed latex particles.

In Figure 4, the electron micrograph shows clearly that the polymer particles are hollow, after the ultrafiltration step allowed to remove the PDMS core. The deformation of the spherical shape of these particles is also visible, and is possibly due to the sonication and subsequent drying on the TEM grid.

Finally, the polymer nanocontainers were labelled with the fluorescent dye, carboxyfluorescein. These dye-loaded nanoparticles were studied in the bulk and on surfaces with a Zeiss 510 upright confocal laser scanning microscope. A 488-nm line of the argon-ion laser was used for the excitation of carboxyfluorescein.

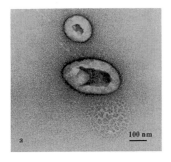 

Figure 4. a/ TEM image of polymer nanocontainers after ultrafiltration in toluene. b/ Confocal image of the labelled nanocontainers.

Figure 4b shows the dye-loaded nanocontainers obtained after several washing cycles to remove the free dye: this result proves clearly the encapsulation of carboxyfluorescein inside the nanocontainers.

Conclusion

In conclusion, the polymer nanocontainers with two different shell thicknesses and various diameters have been synthesized. Next, the morphological study by TEM confirms the formation of spherical particles of a core-shell type. Furthermore, the ultrafiltration step is a well-suited method for removing the core to obtain the polymer nanocontainers. We have also demonstrated that the nanocontainers could be labelled with carboxyfluorescein dye. In the future, the outer surface of nanocontainers will be modified with functional groups that enhance the solubility in aqueous media and allow their covalent binding to surfaces.

[1] W. Meier, *Chem. Soc. Rev.,* **2000**, 5, 295.
[2] F. Caruso, *Adv. Mater.,* **2001**, 13,11.
[3] D. D. Lasic, *Liposomes: from physics to Applications, Elsevier, Amsterdam,* **1993.**
[4] J. D. Morgan, C. A. Jonhson and E. W. Kaler, *Langmuir,* **1997**, 13, 6447.
[5] J. Hotz and W. Meier, *Langmuir,* **1998**, 14, 1031.
[6] H. Huang, E. E. Remsen, T. Kowaleski and K. L. Wooley, *J. Am. Chem. Soc.,* **1999**, 121, 3805.
[7] O. Emmerich, N. Hugenberg, M. Schmidt, S. S. Sheiko, F. Baumann, B. Deubzer, J. Weis and J. Ebenhoch, *Adv. Mater.,* **1999**, 11, 1048.
[8] W. Meier, *Chimia,* **2003**, 56, 490.
[9] M. Sauer and W. Meier, *Colloids and Colloid Assemblies,* **2004**, 150.
[10] T. Sanji, Y. Nakatsuka, S. Ohnishi and H. Sakurai, *Macromolecules,* **2000**, 33, 8524.
[11] X. Z. Kong and E. Ruckenstein, *Journal of Applied Polymer Science,* **1999**, 73, 2235.
[12] X. Zhang, Y. Yang and X. Liu, *Polym. Commun.,* **1982**, 4, 310.
[13] A. De Gunzbourg, J. Favier and P. Hemery, *Polym. Int.,* **1994**, 35, 179.
[14] D. R. Weyenberg, D. E. Findlay, J. Cakata and A. E. Bey, *J. Appl. Polym. Sci,* **1999**, 27, 27.
[15] A. E. Bey, D. R. Weyenberg and L. Seibles, *Polym. Prepr.,* **1970**, 11, 995.

Macromol. Symp. **2005**, *222*, 163-168   163

Preparation and Characterization of Interpenetrating Polymer Hydrogels Based on Poly(acrylic acid) and Poly(vinyl alcohol)

Rebeca Hernández, Daniel López,* Ernesto Pérez, Carmen Mijangos

Instituto de Ciencia y Tecnología de Polímeros, C.S.I.C., Juan de la Cierva 3, 28006 Madrid, Spain
Fax: (34) 91 564 48 53; E-mail: daniel@ictp.csic.es

Summary: Interpenetrating polymer hydrogels (IPHs) of Poly (vinyl alcohol) (PVA) and Poly (acrylic acid) (PAAc) have been prepared by a sequential method: crosslinked PAAc chains were formed in aqueous solution by crosslinking copolymerization of acrylic acid and N, N'-methylenebisacrylamide in the presence of PVA. The application of freezing-thawing cycles (F-T cycles) leads to the formation of a PVA hydrogel within the synthesized PAAc hydrogel. The swelling and the viscoelastic properties of the prepared IPHs were evaluated on the basis of the structural features obtained from solid state ^{13}C-NMR spectroscopy.

Keywords: interpenetrating hydrogels; poly (acrylic acid); poly (vinyl alcohol); solid state ^{13}C-NMR; viscoelastic properties

Introduction

Hydrogels are hydrophilic polymer networks that have a large capacity for absorbing water and that are characterized by the presence of crosslinks, crystalline and amorphous regions, entanglements, and rearrangements of hydrophobic and hydrophilic domains[1]. Polymer hydrogels have been proposed for many applications, such as the controlled delivery of medicinal drugs, artificial muscles, sensor systems, and bioseparations, because of their good biocompatibility, stimuli-responsive properties, and water permeation properties[2]. The reinforcement of a polymer hydrogel is a major problem in the expansion of its applications because a hydrogel has a poor mechanical property in water. Polymers with microphase separated morphologies, such as copolymers in which hydrophobic and hydrophilic domains alternate, seem to possess improved mechanical properties. This morphology can also be achieved with interpenetrating polymer hydrogels (IPHs).

IPHs are a combination of two or more polymers gels synthesized in juxtaposition[3]. They can also be described as polymer gels held together by permanent entanglements. The gels are held by topological bonds, essentially without covalent bonds between them. By

                            DOI: 10.1002/masy.200550420

definition, an IPH structure is obtained when at least one polymer gel is synthesized independently in the immediate presence of another. IPHs are an important class of materials attracting broad interest from both fundamental and applications points of view[4,5].

In the present study we report on the preparation, and on the swelling and viscoelastic properties of temperature, electrical-stimuli and pH-dependent PVA/PAAc interpenetrating hydrogels. The properties of the PVA/PAAc IPHs have been interpreted on the basis of the structural features obtained from solid state ^{13}C-NMR spectroscopy.

Experimental

Acrylic acid monomer was purchased from Aldrich and was purified under vacuum distillation to eliminate hydroquinone inhibitor. N, N´-methylenbisacrilamide (N-BAAm), used as crosslinker and potassium persulfate, used as thermal initiator were employed without further purification. Poly (vinyl alcohol), > 99% hydrolyzed, with a weight average molecular weight of 94.000 g/mol and a tacticity of syndio = 17.2%, hetero = 54.1 % and iso = 28.7%, was from Aldrich and it was used without further purification.

PVA/PAAc IPHs were prepared by a sequential method: PVA solutions (polymer concentrations ranging from 3 to 10 % (g/mL)) were prepared in hermetic Pyrex tubes by mixing the appropriate amount of polymer and water (milli-Q grade) at 100 C under conditions of vigorous stirring until the polymer was completely dissolved. Acrylic acid monomer aqueous solutions containing the thermal initiator, and the crosslinking agent were added at room temperature. The solutions were poured into glass plates, sealed with paraffin and allowed to react at 50 C for 24 hours. The obtained hydrogels were cut into specimens of cylindrical form (20 mm in diameter and 2 mm in height). Some of the specimens were tested in this state and other specimens were subjected to a freezing-thawing cycle: specimens were frozen to -32 C for 15 hours and then, were allowed to thaw at room temperature for 5 h.

Dynamic viscoelastic measurements were performed in a TA Instruments AR1000 Rheometer, using the parallel-plate shear mode to measure the storage modulus, G', the loss modulus, G'' and the loss tangent, tan δ. To avoid the influence of aging on the G' modulus, the measurements for all samples were performed 2 h after the gels were prepared. The operating conditions were the following: temperature sweep between 20 and 100 C, plate diameter 20 mm, frequency 1 Hz, temperature scan 20 C/min, torque 50 µNm. The linear viscoelastic region was located with the aid of a torque sweep. All the viscoelastic measurements were performed on hydrogels swelled to equilibrium.

The cylindrical specimens of the hydrogels were immersed in distilled water and allowed to swell until equilibrium is attained at room temperature. The relative swelling ratio (Q_r) of the samples was defined by equation (1)

$$Q_r = \frac{(W_s - W_r)}{W_r} \tag{1}$$

where W_s and W_r are the weights of the swollen sample and the relaxed sample (specimen as formed, previous to the immersion in distilled water), respectively.

Solid state ^{13}C-NMR experiments were carried out using a Bruker spectrometer at 400 MHz. High resolution ^{13}C NMR was performed using magic-angle sample spinning (MAS) and high power spin decoupling. To enhance the signal to noise ratio, the cross-polarization (CP) technique was applied. Zirconia rotors were used at a spinning velocity of 4.0 KHz. The contact time for CP was 1 ms. 4000 scans were necessary to obtain an adequate signal to noise ratio. The chemical shifts of ^{13}C spectra are reported in ppm relative to TMS by taking the methine carbon of solid adamantane (29.5 ppm) as an external reference standard.

Results and discussion

Figure 1 shows the effect of PVA concentration in the relative swelling ratio of IPHs (crosslinking degree of the PAAc hydrogel 3%). As can be observed, the higher is the concentration of PVA in the IPHs, the lower is the relative swelling ratio in the IPHs subjected to one F-T cycle. In the IPHs not subjected to this treatment the relative swelling ratio decreases until a concentration of PVA of 7% (g/mL). This can be explained by the increase of intermolecular interactions and entanglements density with the increase of polymer concentration. By comparing the swelling ratio of the IPHs subjected and not subjected to one F-T cycle it can be seen that, in the first case, IPHs swell more than in the second case. This could be explained by the physical gelation of PVA within the PAAc network which increases the crosslinking density of the network.

166

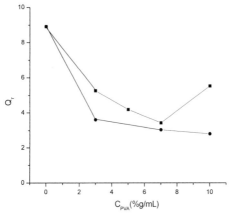

Figure 1. Relative swelling ratio as a function of PVA concentration for PAAc/PVA hydrogels subjected (•) and not subjected (■) to a freezing-thawing cycle. PAAc crosslinking degree 3%.

Concerning the viscoelastic properties, Figure 2 depicts the storage modulus as a function of PVA concentration for IPHs with a degree of crosslinking for the PAAc hydrogel of 3%. As can be seen, the storage modulus of the IPH increases with PVA concentration. Furthermore, the IPHs subjected to one F-T cycle have higher modulus values than IPHs not subjected to this treatment. These results are in accordance with the swelling measurements.

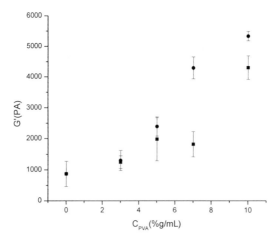

Figure 2. Storage modulus as a function of PVA concentration for PVA/PAAc hydrogels subjected (•) and not subjected (■) to a freezing-thawing cycle. PAAc crosslinking degree 3%.

To obtain basic information on molecular interactions in the PVA/PAAc IPHs, we have performed solid state ^{13}C-NMR spectroscopy of PVA, PAAc and PVA/PAAc IPHs at room temperature. The spectra are shown in Figure 3 and the peak assignment for pure PAAc networks of crosslinking degree 3% and pure PVA films are given in table 1.

Table 1. Peak assignment for PAAc (crosslinking degree 3%) and PVA films.

Peak	Assignment	PAAc 3%	PVA 10% (g/mL)
1	(-CH$_2$-)	34.1	
2	(-CH-)	40.9	
3	(-CH$_2$-)		45.8
4	(-CHOH-)		65.0
5	(-CHOH-)		70.6
6	(-CHOH-)		76.3
7	(-CO(NH)-)	177.7	
8	(-COOH-)	181.8	

Resonance of the COOH carbon of PAAc and CHOH carbon of PVA, whose chemical shifts are very sensitive to hydrogen bond formation, consist of well resolved peaks without any overlapping for all the samples. For pure PVA, the CHOH band shows three peaks in the solid state as previously reported[6]. Terao et al[6] interpreted the chemical shifts of the CHOH resonance in terms of inter and intramolecular hydrogen bonding. They assigned peak 6 to the carbon which is linked by two hydrogen bonds to neighbor CHOH groups, peak 5 to the carbon linked by only one hydrogen bond, and peak 4 to the carbon not hydrogen bonded at all. Thus, peak 6 and 5 can be taken as indicators of inter and intramolecular hydrogen bonding of OH groups between two units of the PVA chain.

In the same way as for PVA/PAAc blends, a remarkable composition dependence of the ^{13}C-NMR spectra was observed for PVA/PAAc IPHs. Peak 6 is not observed for any of the IPHs samples. This detail indicates that the formation of two hydrogen bonds over the same hydroxyl group is restricted. This can be interpreted by the formation a PVA-PAAc complex through hydrogen bonds between the hydroxyl groups of the PVA chains and the carbonyl group of the PAAc chains. The steric hindrance introduced by the PAAc inhibits the formation of more than one hydrogen bond per monomer unit.

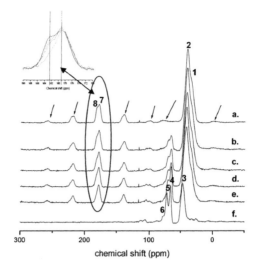

Figure 3. 13C CP/MAS spectra of dried films of: PVA (f), PAAc of crosslinking degree 3% (a) and PVA/PAAc IPHs of PVA concentration 3% (b); 5% (c); 7% (d); 10% (e) and 15% (f).

Conclusion

Interpenetrating polymer hydrogels composed of PVA and PAAc were successfully prepared by a sequential method as suggested by the viscoelastic and the swelling experiments: crosslinked PAAc chains were formed inside of a PVA solution by crosslinking copolymerization of acrylic acid and N,N′-methylenebisacrylamide. Then, the application of freezing-thawing cycles leads to the formation of a PVA hydrogel within the PAAc hydrogel[7]. Besides, solid state ^{13}C-NMR experiments has revealed the existence of hydrogen bonds between PVA and PAAc in IPHs.

[1] C.L. Bell, N.A. Peppas, *Adv. Polym. Sci.* **1995**, 22, 125

[2] C.M. Hassan, N.A. Peppas, *Adv. Polym. Sci.* **2000**, 153, 37

[3] L.H. Sperling, *"Interpenetrating Polymer Networks and Related Materials"*, Plenum Press, New York, 1981

[4] R. Bischoff, S.E. Cray, *Prog. Polym. Sci.* **1999**, 24, 185

[5] R. Yahya, Y. Ahmad, A.W. Mitchell, *Macromolecules* **1999**, 32, 3241

[6] T. Terao, S. Maeda, A. Saika, *Macromolecules* **1983**, 16, 1535

[7] R. Hernández, D. López, C. Mijangos, J.M. Guenet, *Polymer* **2002**, 43, 5661

Effect of Solubility Parameter of Polymer-Solvent Pair on Turbulent Drag Reduction

Chul Am Kim,[1,2] *Hyoung Jin Choi,*[2] Jun Hee Sung,[2] Hyung Min Lee,[2] Myung S. Jhon[3]*

[1] Electronics and Telecommunications Research Institute, Daejon 305-350, Korea

[2] Department of Polymer Science and Engineering, Inha University, Incheon 402-751, Korea
Fax: (+82) 32 865 5178; E-mail: hjchoi@inha.ac.kr

[3] Department of Chemical Engineering, Carnegie Mellon University, Pittsburgh, PA 15213, USA

Summary: Polymer-induced turbulent drag reduction by adding a minute amount of high-molecular weight polyisobutylene (PIB) into two different organic solvents of n-heptane and xylene was examined using a rotating disk system. The dependence of drag reduction (DR) efficiency on various factors such as polymer molecular weight, polymer concentration (C), and solvent quality was examined. Based on the linear relationship between C and C/DR for different molecular weights of PIB, a universal curve was able to characterize a particular polymer/solvent family, independent of the molecular weight of polymer.

Keywords: drag reduction; molecular weight; polyisobutylene; solubility parameter; turbulent flow

Introduction

The drag reduction (DR) phenomenon was originally discovered from the observation of a reduction of the pressure drop in a turbulent pipe flow by an addition of a minute amount of additives, *i. e.,* the wall shear stress or the skin friction drag is drastically reduced by additives.[1] The solvent effect which is related to polymer hydrodynamic volume was found to play an important role on this DR phenomenon. In general, coiled polymer molecules show a different type of drag reduction capability than extended ones.[2, 3] That is, DR efficiency in good solvents has been found to be higher than that in poor solvents.[4] By studying effect of molecular conformation via tuning the interactions between polymer molecules and solvent,[5] it was found that the DR onset point of polystyrene occurred earlier in good solvents (benzene and toluene) than in poor solvents (mixtures of toluene and isooctane). Molecular conformation can also be altered by controlling the salinity of the polyelectrolyte solution.[6] Poly(acrylamide) in a low salt aqueous medium, where the

DOI: 10.1002/masy.200550421

polymer molecules expanded due to charge repulsion, produces more drag reduction than in high salt solutions in which molecules are in coiled state. The molecular conformation of polyacrylic acid (PAA) can be varied by changing the pH value of the solution: at low pH, the polymer molecules collapse due to protonation, while at high pH, the molecules expand due to charge repulsion. The DR ability of PAA was better in high pH than in low pH solutions.[7, 8] Furthermore, DNA has been recently examined as a DR candidate under different pH and salt conditions[2], and its degradation under turbulence was also reported.[9, 10] In this study, we investigated the effects of polymer concentration on DR and characterized its DR efficiency by generalizing the universal curve. The DR efficiency was also correlated with polymer-solvent interaction parameters and the molecular weight of polymer.

Experimental

The polyisobutylenes (PIB) are the Vistanex MM (middle molecular weight) grades acquired from Exxon Mobile Chemical Co. (USA). Four different grades of PIBs based on molecular weight (MM L-80, L-100, L-120, and L-140) were used as drag reducers for organic solvents. PIBs, highly paraffinic hydrocarbon polymers, are composed of terminally unsaturated long straight-chain molecules and have properties of being light in color, odorless, tasteless, and nontoxic. Molecular characteristics of the PIBs are given in Table 1.

Table 1. Properties of PIB.

Polymer grade	Viscosity average molecular weight [a], $\overline{M}_v$ ($\times 10^6$)	Weight average molecular weight [b], $\overline{M}_W$ ($\times 10^6$)
MM L-80	0.99	1.49
MM L-100	1.20	2.67
MM L-120	1.60	3.29
MM L-140	2.10	5.50

[a] Taken from the manufacturer's information
[b] Obtained from GPC (Gel Permeation Chromatography) measurements

n-heptane and xylene (Junsei Chemical Co., Japan) of reagent grade without additional distillation steps were used as solvents performed our DR measurements for PIB with different molecular weights. The density of n-heptane and xylene are 0.684 g/ml and 0.861 g/ml at 20°C and the solubility parameters of these samples are 15.2 $(MPa)^{1/2}$ for n-heptane, 18.0 $(MPa)^{1/2}$ for xylene and 15.5 $(MPa)^{1/2}$ for PIB.

DR measurements were performed using a rotating disk system (RDS), which consists of a stainless steel disk whose dimensions are 14.5 cm diameter × 0.32 cm thickness, enclosed in a cylindrical, thermostatically-controlled container, which is made of a stainless steel disk whose dimensions are 16.3 cm inner diameter × 5.5 cm height. The volume of solution required to fill the entire container is approximately 1,020 cm^3. A DC motor generator coupled with a controller (Cole Parmer Master Servodyne Unit, USA) was used to maintain a preset rotational speed by measuring a variable torque as required by the load on the disk. The fluid temperature and the rotational speed were measured by a K-type thermocouple and a digital tachometer, respectively.

The rotational Reynolds number (N_{Re}) can be defined as $N_{Re} = \rho r^2 \omega / \mu$ in an RDS system. Here, ρ and μ are the fluid density and fluid viscosity, r is a disk radius, and ω is the disk rotational speed. All measurements were performed at the rotational speed of 1,800 rotations per minute (rpm) corresponding to solvent-based N_{Re} of 9.9 × 10^5. The flow becomes turbulent when $N_{Re} > 3 \times 10^5$ or equivalently 1,050 rpm in our RDS system. The drag reduction efficiency was measured first by measuring the torque required to rotate the disk at a given speed in the pure solvent. By measuring the corresponding torque required to attain the same speed in polymer added solution, the percent drag reduction (DR(%)) was then measured as: $DR(\%) = [(T_o - T_p)/T_o] \times 100$. Here, T_o is the torque measured in pure solvent and T_p is the torque measured in polymer solution. Volume of the syringed stock solutions was determined considering the volume of each solvent in the apparatus for each polymer concentration in each stock solution was then carefully syringed into the turbulent flow fields.

Results and Discussion

Polymer induced turbulent DR efficiency generally depends on the nature of the polymer solution. For a given polymer and flow geometry, the DR initially increases with concentration, but eventually becomes independent of concentration. Because the DR is caused by the sum of the contributions from individual polymer molecules, increasing the polymer concentration linearly enhances the DR(%) effect initially. Figure 1 shows the DR of four different molecular weights of PIB in n-hepatne. The obtained optimum concentration exhibiting the maximum DR decreases as the molecular weight of the PIB increases. The values of the maximum DR for different molecular weights of PIB are used in the analysis of universal curves.

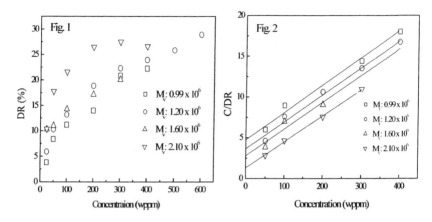

Figure 1. Percent DR *vs.* C for four different molecular weights of PIB in n-heptane.
Figure 2. C/DR *vs.* C for four different molecular weights of PIB in n-heptane at 1,800 rpm.

The $DR_{sp} = DR/C$ at infinite dilution gives the intrinsic DR, $[DR] = \lim_{C \to 0} DR/C$, which is a measure of drag reduction efficiency for an initial increment of the polymer molecule. In the turbulent DR for known polymer and flow rate, DR_{sp} monotonically decreases with an increase in C. The DR_{sp} *vs.* C curve possesses a characteristic shape presenting straight-line asymptotes $DR_{max} = \lim_{C \to \infty} DR$, where DR_{max} is the maximum DR(%) given in Fig. 1. The limit $C \to \infty$ leads to concentrated solutions in a thermodynamic sense. Before establishing an empirical relationship, we define the intrinsic concentration as follows: $[C] = DR_{max}/[DR]$. With the same number of parameters[11] and the Padé form fits the DR data much better than a Taylor expansion.

$$DR = \frac{C[DR]}{K + C/[C]} \text{ or } \frac{DR/C}{[DR]} = \frac{1}{K + C/[C]}. \tag{1}$$

The two important parameters, [DR] and [C], are previously reported and found that more efficient drag reducers have a larger DR_{max} and a smaller $[C]$[5] for various polymer-solvent systems. To analyze the data, Eq. (1) is written in the following form:

$$\frac{C}{DR} = \frac{K[C]}{DR_{max}} + \frac{C}{DR_{max}}$$

(2)

Equation (2) implies that there is a linear relationship between C/DR and C up to the optimum concentration of each molecular weight. The linear correlations between C/DR and the polymer concentration for four different molecular weights of PIB in n-heptane for a range of conditions close to the maximum drag reduction are given in Fig. 2. The development of the empirical function to relate DR to relevant solution properties, *i.e.*, concentration, successfully describes the drag reduction results up to concentrations somewhat below that needed to produce an optimum DR.[12] Therefore, [C] is an extremely useful quantity in normalizing the drag reduction data of different molecular weight compounds in a homologous series. Figures 3 shows the plot of DR(%) *vs.* C/[C] for PIB-heptane system.

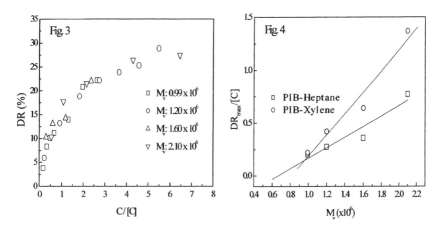

Figure 3. DR(%) *vs.* C/[C] of PIB in n-heptane at 1,800 rpm.
Figure 4. $DR_{max}/[C]$ index as a function of viscosity average molecular weight for PIB in n-heptane and xylene at 1,800 rpm.

The parameter $DR_{max}/[C]$ from the Henry's law defines the efficiency of the polymer additives on a unit concentration basis (at the infinite dilution limit):

$$\lim_{C \to 0} \frac{DR}{C} = \lim_{C \to 0} \frac{DR_{max}}{C + [C]} = \frac{DR_{max}}{[C]}.$$

(3)

Figure 4 shows correlations between $DR_{max}/[C]$ *vs.* M_v for the PIB family in n-heptane and xylene. Each plot is linear and the intercepts are $M_v = 0.67 \times 10^6$ for PIB in n-heptane and $M_v = 0.78 \times 10^6$ for PIB in xylene in which no DR effect takes place at the corresponding Reynolds number.

Thereby, Eq. (2) is regarded as the universal curve for polymer-induced turbulent drag reduction. If we define $(DR/C)/[DR]$ as β and $C/[C]$ as α, the correlation is expressed by $\beta = 1/(K + \alpha)$. For a PIB-toluene in a capillary tube system[13] the universal curve was obtained as $\beta = 1/(0.4 + \alpha)$. From these results, the constant K becomes a parameter inherent to the interaction between a polymer and a solvent. Therefore, the constant K is thought to be a characteristic of a particular polymer family in a given solvent and does not depend on the molecular weight and flow geometry. In other words, the constant K is characteristic of a particular polymer-solvent pair, such that K=0.8 for PIB-heptane and 1.1 for PIB-xylene in this work. The lower K value implies the better solvent with a higher DR efficiency.[14]

Conclusion

The effects of concentration and molecular weight on DR for a homologous series of PIB with different solvents were investigated using RDS. The better solvent produces a higher DR effect under the same conditions. Using the universal curve, it was found that K values for different polymer and solvent pairs were proportional to the solvent quality and better solvents yield lower K values.

Acknowledgement

This study was supported by research grants from the Korea Science and Engineering Foundation through the Applied Rheology Center at Korea University, Seoul, Korea.

[1] M. S. Jhon, G. Sekhon, R. Armstrong, *Adv. Chem. Phys.*, **1987**, 66, 153.
[2] C. Wagner, Y. Amarouchene, P. Doyle, D. Bonn, *Europhys. Lett.*, **2003**, 64, 823.
[3] G. Boffetta, A. Celani, S. Musacchio, *Phys. Rev. Lett.*, **2003**, 91, 034501.
[4] R. C. Little, R. L. Patterson, *J. Appl. Polym. Sci.*, **1974**, 18, 1529.
[5] O.K. Kim, T. Long, F. Brown, *Polym. Commun.*, **1986**, 27, 71.
[6] P. S. Virk, *AIChE J.*, **1975**, 21, 625.
[7] C. A. Parker, A. H. Hedley, *Nature Phys. Sci.*, **1972**, 236, 61.
[8] S. H. Banijamali, E. W. Merrill, K. A. Smith, L. H. Peebles, *AIChE J.*, **1974**, 20, 824.
[9] H. J. Choi, S. T. Lim, P. K. Lai, C. K. Chan, *Phys. Rev. Lett.*, **2002**, 89, 088302.
[10] S. T. Lim, H. J. Choi, S. Y. Lee, J. S. So, C. K. Chan, *Macromolecules*, **2003**, 36, 5348.
[11] H. J. Choi, M. S. Jhon, *Ind. Eng. Chem. Res.*, **1996**, 35, 2993.
[12] C. A. Kim, K. Lee, H. J. Choi, C. B. Kim, K. Y. Kim, M. S. Jhon, *J. Macromol. Sci.*, **1997**, A34, 705.
[13] E. Dschagarowa, G. Mennig, *Rheol. Acta.*, **1977**, 16, 309.
[14] J. I. Sohn, C. A. Kim, H. J. Choi, M. S. Jhon, *Carbohydrate Polym.*, **2001**, 45, 61.

Macromol. Symp. **2005**, *222*, 175-180

Thermoreversible Gelation of Poly(vinylidene fluoride) – Camphor System

D. Dasgupta,[1] Swarup Manna,[1] Sudip Malik,[2] Cyrille Rochas,[3] Jean Michel Guenet,*[2] A. K. Nandi*[1]

[1] Polymer Science Unit, Indian Association for the Cultivation of Science, Jadavpur, Kolkata – 700 032, India
E-mail: psuakn@mahendra.iacs.res.in
[2] Institut Charles Sadron, CNRS UPR22, 6, rue Boussingault, 67083 Strasbourg, Cedex, France
[3] Laboratoire de Spectrométrie Physique CNRS-UJF UMR5588, 38402 Saint Martin d'Hères Cedex, France

Summary: Poly (vinylidene fluoride) (PVF$_2$) produces thermoreversible gel in camphor when quenched to 25^0C from the melt under sealed condition. The SEM micrograph of dried PVF$_2$/camphor gel (W$_{PVF_2}$ = 0.25) indicates presence of fibrillar network structure and the gels at different composition shows reversible first order phase transition. The phase diagram of the gel suggest the formation of a polymer- solvent complex. The melting enthalpy gives a stoichiometric composition of the complex at W$_{PVF_2}$ = 0.25. This corresponds to a molar ratio of *PVF$_2$ monomer/camphor* $\approx$4/5. Temperature-dependent synchroton experiments further support the conclusions derived from the phase diagram.

Keywords: camphor; phase diagram; polymer solvent complex; poly (vinylidene fluoride); thermoreversible gel

Introduction

Thermoreversible gels have been extensively studied for the last two decades [1-3]. Poly(vinylidene fluoride) (PVF$_2$) is a technologically important polymer and produces thermoreversible gels in different solvents containing >C=O group such as diesters [4-7]. Extensive studies of the physical properties of the PVF$_2$ gels were made for diesters of varying intermittent length [7]. The thermoreversible PVF$_2$ gels display a three dimensional fibrillar network structure. Keeping this network structure unaltered is important in dried gels that can be used as porous material. However, the drying of these gels while keeping intact the original network structure is a difficult task. In this aim, we have studied PVF$_2$/camphor systems, that produce thermoreversible gels. The advantage of camphor lies in its propensity of sublimating easily so that solvent removal occurs under near "freeze-drying" conditions. In this short paper we report on the morphology, the thermodynamics (phase diagram) and the crystalline structures of these gels.

   DOI: 10.1002/masy.200550422

Experimental

Poly(vinylidene fluoride) (PVF$_2$) is a product of Aldrich Chemical Company Inc. The weight average molecular weight ($\overline{M}_w$) of the sample is 180 000 g/mol and polydispersity index is 2.54 as obtained from GPC. The PVF$_2$ sample was recrystallized from its 0.2% (W/V) solution in acetophenone. Camphor was purified by sublimation procedure.

The PVF$_2$ and camphor were taken in a thick walled glass tube (8 mm in diameter) and were sealed under vacuum(10^{-3} mm Hg). The sealed tubes were melted at 210 ^{0}C in an oven for 20 min with intermittent shaking to make homogeneous. They were then quenched to 30 ^{0}C. For SEM study the samples were taken out from the tube and kept in a vacuum for 2 days. It was then gold coated and was then observed in a SEM apparatus (Hitachi S-415 A). For temperature dependent X ray investigation gels were prepared by melt quenched method as above and were then cut into circular sample of diameter 4 mm and placed into an equal size window of an aluminum holder, both sides were glued with mica of thickness 25 micron to prevent solvent evaporation. The data were taken at the Synchroton X-ray radiation facility (ESRF) at Grenoble, France, on D2AM beam line.

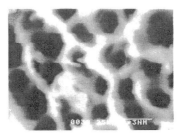

Figure 1. SEM picture of dried PVF$_2$- camphor gel (W$_{PVF_2}$ =0.25).

For thermodynamic study the gels were prepared in Perkin Elmer large volume capsules (LVC) by taking appropriate amount of polymer and camphor and the capsules were tightly sealed with help of a quick press. The samples were subsequently made homogeneous by keeping at 200 ^{0}C in DSC for 15 min with occasional shaking, and then were gelled at –30 ^{0}C for 10 min by cooling from 200 ^{0}C at the rate of 200 ^{0}C / min. They were then heated at the scan rate of 40 ^{0}C/min, the higher heating rate was chosen to avoid recrystallization processes. The enthalpies and melting temperatures were determined by means of a computer attached to the instrument.

Result and discussion

In Fig-1 the SEM picture of dried PVF_2/camphor gel ($W_{PVF_2} = 0.25$) is shown. From the figure fibrillar network structures are clearly observed. The heating thermograms (Fig- 2) of the gel exhibit *first order* phase transitions. Thus the presence of fibrillar network structure and first order phase transitions indicate formation of thermoreversible gel [2]. In figure 3 are presented the temperature-concentration phase diagram as obtained from the maxima of the 1st order transitions together with the Tamman's diagram (enthalpies vs concentration). Two non-variant temperature events can be observed at T= 135°C and T= 150°C. Also, one observes that the 1st order transition in the camphor-reach systems occur at temperatures higher than the pure camphor melting point. We conclude that the 1st order transition is connected to a liquid-liquid phase separation.

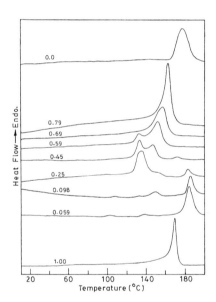

Figure 2. DSC traces obtained at 40°C/mn for variation PVF_2/camphor compositions (as indicated).

Applying GIBBS rules for establishing the phase diagram, we conclude that the first non-variant event corresponds to the incongruent-melting of a compound (C_1) while the second one corresponds to the formation of an eutectic between a camphor solid solution (S_c) and pure PVF$_2$. Note that the composition of the camphor solid solution (S$_c$) varies with temperature. It contains little PVF$_2$ at low temperature but more as temperature is increased. The variation of the enthalpies are consistent with the present description of the phase diagram. Note that in the PVF$_2$-rich systems, the enthalpies associated with the liquidus line do not vary linearly as they should. This arise from the fact that the maximum at 150°C representing the eutectic melting event cannot be properly extracted from the DSC traces. If we consider that the enthalpy associated with the liquidus should vary linearly (dotted line) then we can calculate what should be the values for the enthalpy of the non-variant event at 150°C. These values do fall on the dashed line expected for the variation of the enthalpies associated with the eutectic melting.

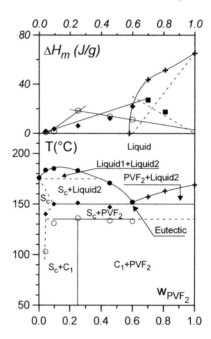

Figure 3. Phase diagram of PVF$_2$/camphor systems obtained from DSC melting endotherms. As is customary, the dotted lines stand for event that are not explicitly observed, but that should exist on the basis of GIBBS phase rules.

From the maximum at W$_{PVF_2}$ = 0.25 of the enthalpy associated with the event at T= 135°C (incongruent-melting of the compound) we deduced a stoichiometric composition for C$_1$ of *4 PVF$_2$ monomer unit/5 camphor molecules.*

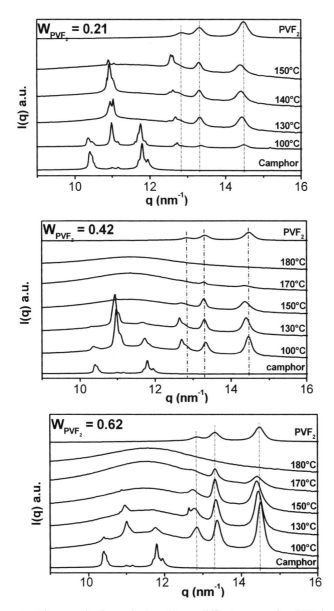

Figure 4. Time-resolved synchroton X-ray diffractograms for *PVF₂/camphor* systems. Compositions and temperatures as indicated. Dotted lines show the position of the reflections for pure PVF₂. Here, the heating rate was about 2°C/mn.

Time-resolved X ray patterns for some of PVF_2 /camphor compositions are shown in Fig-4. together with the diffractograms at 20^0C of pure camphor and pure PVF_2.

Some noticeable alterations of the X-ray patterns have occurred compared to the linear combination of the components diffractograms. Note that a shift of the transitions is expected with respect to those reported on phase diagram because this latter is drawn on the basis of the maxima of the endotherms while the X-rays patterns, particularly the disappearance of reflections, are linked to the end of the endotherm.

In the three spectra shown in figure 4 one can observe a strong reflection at $q \approx 11$ nm^{-1} which disappears above 140°C. It does not correspond to any of the reflections of either pure camphor nor pure PVF_2. We suggest that this peak stands for the crystalline lattice of compound C_1. The solid solution S_c is most probably characterized by the peaks at $q \approx 12.6$ nm^{-1}, together with those of pure PVF_2. It is likely that this phase consists of camphor crystals alternating with PVF_2 crystals in a way already described by Wittmann and St John Manley [8]. It is also worth emphasizing that the composition of the solid phase S_c is temperature-dependant. Below 135°C, camphor crystals are probably large enough so as to give the camphor reflections at 10.4 and 11.9 nm^{-1}. While above 135°C, the situation changes, hence the disappearance of these reflections.

Conclusion

PVF_2 produces thermoreversible gels with camphor having fibrillar network morphology and reversible first order phase transition. Evidence for a polymer-solvent compound in this gel has been put forward from both thermodynamic study and synchroton X-ray study of the samples.

Acknowledgement

We gratefully acknowledge the financial assistance from the *Indo-French Centre for the Promotion of Advanced Research* (IFCPAR, New Delhi, Grant no. 2808–2).

ESRF (Grenoble, France) is also acknowledged for providing access to the D2AM beam line.

[1] P. S. Russo, Ed. *Reversible polymeric gels and related systems*; ACS Symp. Ser.: New York, **1986**.
[2] J. M. Guenet *Thermoreversible gelation of polymers and biopolymers*; Academic Press: London, **1992**.
[3] K. te. Nijenhuis *Adv. Polym. Sci* **1997**, *130*.
[4] S. Mal; P. Maiti; A. K. Nandi *Macromolecules* **1995**, *28*, 2371.
[5] S. Mal; A. K. Nandi *Langmuir* **1998**, *14*, 2238.
[6] A. K. Dikshit, A. K. Nandi *Macromolecules* **1998**, *31*, 8886.
[7] A. K. Dikshit; A. K. Nandi *Macromolecules* **2000**, *33*, 2616.
[8] J.C. Wittmann; R. St John Manley *J. Pol. Sci. Polym. Phys. Ed.* **1977**, *15*, 1089.

Thermodynamics of Swelling of Poly(N-vinylpyrrolidone)-Poly(ethylene glycol) Complex Due to Water Sorption

Olga Krasilnikova,* Ruben Vartapetian

Institute of Physical Chemistry, Russian Academy of Sciences, Leninsky Prospect 31, Moscow 119991, Russia
E-mail: volosh@phyche.ac.ru

Summary: The deformation of sorbent caused by the sorption is new method of quantitative investigation "in situ" of interaction in system host-quest. The deformation of PVP-PEG complex, $\varphi_{PEG}=0.36$ and $\varphi_{PEG}=0.20$ due to water sorption has been studied by the measuring of the relative elongation of the polymer samples and the isotherms of water sorption simultaneously. The investigation of the sorption deformation gives the possibility of direct estimation of polymer sample free volume and it's variation during sorption, also the variation of Gibbs energy of system due to sorption according to the vacancy solution theory. The glassy-plastic state transition of polymer during water sorption has been observed.

Keywords: glass transition; polymer free volume; poly(N-vinyl pyrrolidone)-poly(ethylene glycol) blends; thermodynamics of swelling of polymer; water sorption

Introduction

Thermodynamic parameters of water sorption by polymer blends - of poly(N-vinyl pyrrolidone)-poly(ethylene glycol)(PVP-PEG) were investigated using the method of quantitative estimation of interaction in the system host-quest "in situ" by the studies of polymer swelling in water vapours. The interaction of water molecules (sorbate-solvent) with atoms of polymer (sorbent) leads to variation in bond lengths of polymer molecules and the volume of polymer is changed. There are the expansion and contraction of every sorbents, especially polymers due to sorption process. PEG molecules bonded to comparatively long PVP chains are forming the supramolecular gel network [1]. The interaction of elastic chains of PEG with chains of PVP forms the free volume of PVP. The sorption of water leads to variation in volume of the polymer blend. The sorption, thermodynamic properties and free volume of supramolecular structure of PVP-PEG blends depend on the composition of blend, water content and thermal treatment. The formalized theoretical treatment of isotherms based on the assumption of the inert character of sorbent has usually ignored the existence of variation of sorbent volume, as well as the variation of energy of sorbent during the sorption and is not appropriate for analysis of sorption of swelling polymers. The phenomenon of sorption deformation has been interpreted

DOI: 10.1002/masy.200550423

using the model of vacancy solution [2]. Both the contraction and the expansion of sorbent due to sorption are consistently explained by the special thermodynamic theory, based on the model of binary vacancy solution. The theory connects the sorption deformation and Gibbs potential of two component system sorbent-sorbate by a relationship. The study of sorption deformation allows to elaborate the thermodynamic function of sorption process which takes into account the variation of the chemical potential of sorbate, as well as of sorbent . That gives the possibility of direct estimation of free volume of polymer sample and it's variation during sorption, as well as the variation of Gibbs energy of system due to sorption.

The present communication deals with the investigation of variation of volume of sorbents - PVP-PEG blends - during water sorption, and demonstrates the possibilities of this method for thermodynamic description of sorption and swelling process, the variation of free volume of system polymer-water, the transition of polymer into plastic state caused by sorption.

Experimental

The deformation of polymer blends PVP-PEG, $\varphi_{PEG}=0.36$ and $\varphi_{PEG}=0,20$ due to water sorption has been studied by the simultaneous measuring of the variations of linear dimensions (relative elongation and contraction) of the polymer sample and the isotherms (at 20 ^{0}C) of water sorption after the thermal and vacuum treatment of polymer sample. Poly(N-vinyl pyrrolidone) $M_w=10^6$ g/mol, Poly(ethylene glycol) with $M_w=400$ g/mol were mixed to obtain the blend with 36% and 20% PEG. These blends were dried at 100 °C after that the pellet of polymer (length 30 mm) was placed into the vacuum automatic microbalance with sensitivity 10^{-5} g. The variations of linear dimensions (relative elongation or contraction) of the sample placed into microbalance were measured with an accuracy of 0,05% during the sorption process. Before the water sorption the sample was outgassed in vacuum up to constant weight at 20 ^{0}C . After water sorption-desorption experiment the sample was outgassed at 100 ^{0}C, and at low filling the second and the third sorption- desorption experiments were performed for PVP-PEG-36 blend.

Results and discussion

The sorption, sorption deformation, thermodynamic properties and free volume of supramolecular structure of PVP-PEG blends depend on the composition of blend, the water content and thermal and vacuum treatment of polymer blend before the investigation.The experimental data obtained are given as the value of relative elongation of the pellet $\Delta L/L_0$ on the sorption value (Figure 1). The thermal treatment of polymer leads both to the decrease of free volume of polymer and the free energy of system.

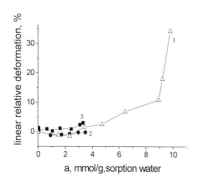
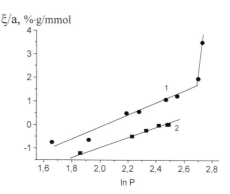

Figure 1. The deformation of PVP-PEG-36% caused by water sorption: 1- after outgassing, 2- heat treatment, 3-two heat treatments at 100 °C .

Figure 2. The ratio of relative volume deformation ξ to sorption a as a function of lnP : 1- outgassing, 2 – heat treatment (according to Equation 3).

The equilibrium dependence of variation in linear dimensions of pellet of PVP-PEG-36 blend on sorption shows the significant (up to -1,5 %) contraction at small water filling (Fig.1, curve 1). At higher fillings PVP-PEG complex expanded up to 35%. After thermal outgassing at 100 °C the contraction of polymer at low water filling is diminished (curves 2, 3 showed at low filling). The expansion of PVP-PEG-36 is more than that of PVP-PEG-20, but the contractions at low filling of these polymer blends are near 1,5%.

The contraction of the polymer sample at low water fillings may be associated with the sorption of water into hollow voids in polymer structure. It may be assumed, that this voids are similar to micropores. They were generated due to drying and thermal treatment. It was suggested, that the volume of this voids is polymer free volume. The water molecules interact with two opposite walls of micropore, and the polymer is contracted. The phenomenon of contraction of the sorbent due to sorption has been demonstrated by many careful studies [3], but the formalised theoretical treatments of sorption have usually ignored the existence of this deformation as well as the variation of energy in system caused by this deformation. The estimation of free energy of sorbent- sorbate system based only on the analysis of the isotherms of sorption has neglected the variation of energy of sorbent due to sorption. According to the investigation of sorption deformation the assumption of inert character of sorbent for polymer is mistaken. The phenomenon of sorption deformation has been interpreted by the special thermodynamic theory of vacancy solutions, which allows the indirect characterisation of the contraction of sorbent in low filling.

The deformation caused by the sorption is connected with variation of energy of binary system sorbent-sorbat according to the thermodynamic of vacancy solutions [2]. The deformation caused by sorption is determined by depends on the variation of the free energy Gibbs of binary system:

$$\xi = \Delta V/V_0 = \Delta G/K \ V_m = \Delta G \ a_m/KW_0 \qquad (1)$$

where V - the sorbent volume, V_0 - initial volume, V_m - partial molar volume of vacancy solution, W_0 - the summary pore volume, a_m - saturation sorption, K - modulus of elasticity, ΔG - the variation of free energy Gibbs of binary vacancy solution. The vacancy solution exists in osmotic equilibrium with free vacancies, and the chemical potential of vacancy in vacancy solution is zero [2]. According the model of vacancy solution, in equilibrium state chemical potential of sorbate in gas phase and in the vacancy solution are equal. Then the variation of chemical free energy Gibbs is:

$$\Delta G = x\mu + x^*\mu^* = x\mu = x\mu_g = (\mu_0^g + RT\ln P) \ a/ \ a_m \qquad (2)$$

with $x = a/a_m$ - molar fraction of sorbate and x^* - molar fraction of vacancy in vacancy solution and μ and μ^* - chemical potentials of sorbate and vacancies respectively. Substitution of Eq.1 with ΔG into Eq. 2 gives the expression for the deformation isotherm Eq.3:

$$\xi/a = \Delta G/ \ a \ K \ V_m = B + RT/KW_0 \ \ln P \qquad (3)$$

Here B is the constant dependent on the temperature and the standard chemical potential of sorbate in vacancy solution, R - gas constant and T - temperature. The experimental data were fitted with Eq. (3) as shown in Figure 2. The Eq.3 is the straight line in the interval of sorption value up to 8 mmol/g. For the higher sorption values the expansion is so large, that this phenomenon may be considered as the transition of glassy state of structure to plastic state. This dependence (3) has two parts, that indicates on the sharp transition of their thermodynamic properties. The modulus of elasticity for glassy state obtained by Eq.3 is $K = 1.3*10^7$ Pa. This value is adjusted with modulus of elasticity of N,N '-diethyl acrylamide - water, obtained by this method previously [4].

Furthermore the solution thermodynamic analysis of the system performed to Eq.(3) shows the similarity the sorption deformation curves and curves of the variation of Gibbs energy of polymer-water system, provided the linearity of the Eq.3 curve.

The measurements of the sorption deformation of polymer sorbent such as PVP-PEG blend gives the direct estimation of free volume of polymer sorbent and it's variation during the sorption. The curves of sorption deformation (Figure 1) demonstrate the contraction range. It may be assumed, that the volume contraction of -4 % corresponds to the filling of initial free volume of polymer sorbent - 0.0431 cm^3/g. After that the polymer expands up to 0.977 cm^3/g. During the sorption process the partial molar volume of sorbent is changed in accordance with the curve of sorption deformation. The heat treatment in vacuum at 100 ^{0}C leads to the increasing of sorption deformation. The second heat treatment leads to the diminishing of initial contraction and the maximum swelling volume due to the ageing of polymer (curve 3, Fig.1). The ageing of PVP-PEG blend was corroborated by NMR PFG measurements of self-diffusion [1].

Conclusion

The new method of quantitative estimation of interaction in system host-quest "in situ" by the studies of polymer swelling in water vapours gives the possibility of direct estimation of the variation of Gibbs energy of system sorbent-sorbate due to sorption, free volume of polymer sample and it's variation during sorption. According to the special thermodynamic theory of vacancy solutions the deformation of sorbent due to sorption is caused by the Gibbs potential of binary component sorbent-sorbate system. The variation in volume of PVP-PEG (36% and 20%) polymer blends caused by water sorption and isotherms of sorption were measured with and without the heat treatment of the sample. The contraction of sorbent at low water fillings gives the direct estimation of free volume of polymer and it's variation during sorption. The effect of thermal treatment upon the sorption deformation was shown. The modulus of elasticity due to water sorption was obtained for glassy state. The glassy-plastic state transition of polymer during water sorption was observed.

[1] R.Sh. Vartapetian, E.V. Khozina, J. Karger, D. Geschke, F. Rittig, M. Feldstein, A.Chalych, *Macromol.Chem.Phys.* **2001**, 202, 2648.
[2] B.P. Bering , O.K. Krasilnikova, V.V. Serpinski, *Bull.of Acad.Sci of the SSSR,* **1978**,v.27,N 12, 2515 (Eng.)
[3] A.I. Sarakhov, V.F. Kononyuk, M.M. Dubinin . in *"Molecular Sieves"*, *ADVAN.CHEM.SER*, **1973**, 121,p.403
[4] O.K.Krasilnikova, M.E. Sarylova, A.V. Volkov, L.I. Valuev, I.V. Obydennova, *Vysokomol. Soedin..Ser.B.,***1991**, vol.32, no.3, 202 (in Russian).

Macromol. Symp. **2005**, *222*, 187-194

Simulation of Electrical and Thermal Behavior of Poly(propylene) / Carbon Filler Conductive Polymer Composites

Jean François Feller,[*1] *Patrick Glouannec,*[2] *Patrick Salagnac,*[2] *Guillaume Droval,*[1,2] *Philippe Chauvelon*[2]

[1] Laboratory of Polymers, Properties at Interfaces & Composites (L2PIC), Centre de Recherche, BP 92116, 56321 Lorient Cedex, France
E-mail: jean-francois.feller@univ-ubs.fr
[2] Laboratory of Thermal Studies, Energetics & Environment (LET2E), Centre de Recherche, BP 92116, 56321 Lorient Cedex, France

Summary: PP-carbon CPC show interesting thermo-electrical properties, smooth resistivity increase with temperature up to 150°C and consequently high power dissipation on a wide temperature range. The addition of short carbon fibers to PP already formulated with carbon black increases sharply the electrical conductivity of the CPC but does not have much influence on thermal conductivity as it could have been expected from the favorable aspect ratio of the fibers. The simulations of the thermo-electrical behavior of the CPC under tension put into evidence a temperature gradient at high heat flux due to the low thermal conductivity, which may damage the material itself.

Keywords: conductive polymer composite; conductivity models; fillers; simulations; thermal and electrical properties

Introduction

Conductive polymer composites (CPC) resulting from the association of an insulating polymer matrix with electrical conductive fillers (carbon black, carbon fibers, metal particles), exhibit several interesting features due to their electrical resistivity variation with thermal solicitations, the so called PTC effect (positive temperature coefficient) [1]. Besides the numerous advantages of CPC, their low thermal conductivity resulting from their high insulating polymer content is a drawback to achieve high thermal efficiency and allow good heat dissipation. Indeed, the crossing of electric direct current through the charged polymer induces internal power dissipation by joule effect, which can generate temperature gradients very important in the material. The objective of this study is that way to quantify the effect of different fillers and to understand the different phenomena involved during ohmic heating with CPC. In this paper, the results obtained with Poly(propylene) containing carbon black and carbon fibers are presented.

In a first step, the effect of filler content on electrical and thermal conductivities is quantified. Then, the coupling of electrical and thermal phenomena is studied.

© 2005 WILEY-VCH Verlag GmbH & KGaA, Weinheim DOI: 10.1002/masy.200550424

Experiments of ohmic heating are performed on extruded tapes samples. The objective is to validate the values obtained for the thermo-physical parameters and to well understand the material behavior when it is subjected to electrical potential difference as it is the case in ohmic heating.

Materials and methods

Materials. The desired CPC composition was adjusted blending a poly(propylene) master batch containing 30% w/w special P carbon black from PREMIX (TP 6504) with a pure poly(propylene) from TOTAL (3181). THORAYCA T300 long carbon fibers sized with an epoxy agent were obtained from THORAY and cut by APPLY CARBON to obtain short carbon fibers (SCF) with a mean diameter of 8.7 ± 1.7 µm and a length distribution of 1700 ± 0.3 µm. Carbon black filled poly(ethylene-co-ethyl acrylate) (EEA-CB) was LE 7704 from BOREALIS. The main thermal characteristics of the polymers are recalled in Table 1.

Blending. Depending on the sample thickness required for each characterization technique, the blending process chosen was either a BRABENDER LAB STATION internal mixer to obtain 10 mm thick molded pads, or a BRABENDER LAB STATION twin screw extruder to obtain 3 mm thick tapes. To obtain a good dispersion of the SCF in the CPC a two step process was used: a first blend in a twin screw extruder followed by a second in a single screw extruder (after grinding).

Table 1: Polymers characteristics.

	T_g (°C)	$T_{c,n}$ (°C)	T_m (°C)	k (W.m^{-1}.K^{-1})	ρ (Ω.cm)	X (%)
PP	- 10 ± 3	114.4 ± 0.5	166.4 ± 0.5	0.24	10^{11}	52.5
EEA	- 33 ± 3	83 ± 0.5	99.5 ± 0.5	0.26	10^{15}	20

Methods. Electrical resistivity evolution with temperature was measured by a four probe technique. After connection the sample (2 × 10 × 70 mm^3) was placed in an oven [2]. The specific heat capacity C_p (J.kg^{-1}.K^{-1}) was measured using a METTLER TOLEDO DSC 822 differential scanning calorimeter. A three-curve analysis method was used, with sapphire as reference material. The density d was determined with a pycnometer at room temperature and the density variation as a function of temperature was obtained by a TA Instrument DMA 2980 dynamic mechanical analyzer with the compression accessory.

Thermal diffusivity α was determined by the laser flash method. Thermal conductivity k as a function of temperature was deduced from the parameters (α, d, C_p) using equation 1.

$$k = \alpha \, d \, C_p \qquad (1)$$

The effective thermal conductivity as a function of filler content was measured by a guarded hot plate method at 50°C. The samples were 10 mm thick pads of 60 × 60 mm^2 section.

Electrical and thermal properties

Electrical properties. Figure 1 shows that an important resistivity decrease can be obtained by addition of carbon black up to 17.5% v/v CB. Over this content, the curve reaches a plateau and introducing more CB in PP won't decrease the CPC resistivity, but adding carbon fibers up to 5% v/v allows to overcome this limit and to gain 0.75 decade of conductivity.

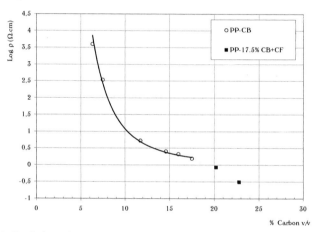

Figure 1. Evolution of PP resistivity with CB content and SCF content over 17.5% v/v CB.

In Figure 2 the variation of the electrical resistivity of PP CPC for three filler contents (30% w/w CB, 30% w/w CB + 5% w/w SCF and 30% w/w CB + 10% w/w SCF) is compared to that of EEA CPC with 27.75% w/w CB. Very different electrical behaviors can be observed: PP CPC resistivity increases smoothly from 40 to 160°C of about one decade while for EEA CPC, a stronger PTC effect is observed (about 2.5 decades between 40 and 100°C). More, the room temperature resistivity of the former is much lower, $\rho \# 0.3 \, \Omega.cm$ for PP + 30%CB + 10%SCF compared to EEA + 27.75%CB, $\rho \# 215 \, \Omega.cm$. This important difference can mainly be explained by the dispersion level of the filler in the matrix, a good dispersion of the filler leads to a lower conductivity, and by the structure of the filler, a highly structured filler is more conductive for the same content. Related the melting of the matrix, the switching temperature is observed at about 80°C for EEA CPC and suggested to be at about 160°C for PP CPC.

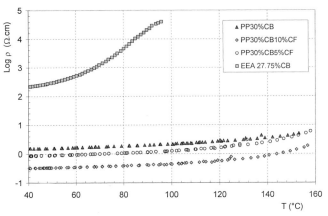

Figure 2. Evolution of electrical resistivity versus temperature.

Thermal properties. The influence of fillers on the thermal conductivity was first investigated (Figure 3). It can first be noticed that the two CPC have similar evolutions of thermal conductivity whatever the filler nature. It can be seen that the effective thermal conductivity of CPC k can only be increased by a factor two over 25% v/v of fillers. On the other hand using higher filler contents unexpectedly also leads to an important modification of rheological properties of CPC, so that it is not realistic to use more than 30% v/v of filler to increase thermal conductivity. Several empirical models predicting the composite effective thermal conductivity k as a function of the filler volume content ϕ for known thermal conductivities of the matrix and the filler can be found in the literature [3]. A very good fitting of experimental data is obtained with MAXWELL [4] volume fraction for randomly distributed and non-interacting homogeneous spheres in a homogeneous medium. Some experimental considerations are taken into account in the LEWIS & NIELSEN model, through parameters linked to particles geometry and the maximum packing volume fraction. The fitting of experimental data with this later model provides two coefficients A = 1.5 and ϕ_m = 0.637 [5] suggesting a random distribution of spherical shape. This result shows that introducing up to 5% v/v of cylindrical elements in the CPC does not modify the overall thermal conduction mode.

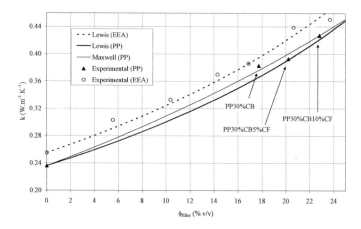

Figure 3. Effective thermal conductivity as a function of carbon volume fraction at 50°C.

Figure 4 represents the evolution of equivalent volume heat capacity (d C_p) and the thermal conductivity with temperature for PP-30% CB. As PP crystals melt over 120°C, a rough increase C_p is observed, this is why C_p has only been evaluated in the temperature range [20°C - 120°C] where the curve is approximately linear. The evolution of the thermal conductivity with temperature was derived from equation (1). The k value obtained at 50°C is in good agreement with that measured independently by the guarded hot plate method.

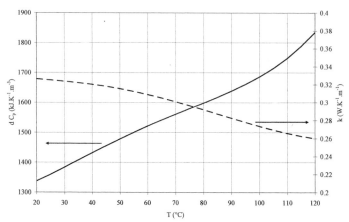

Figure 4. Thermal conductivity and volume heat capacity evolution with temperature for PP-30% w/w CB.

Simulation of electrical and thermal behavior of the CPC

Experimental and modeling work has been carried out on PP-30% CB extruded tapes subjected to direct current, as is the case during ohmic heating. Samples were extruded tapes (dimensions: 179 × 40 × 2.35 mm). They are disposed vertically on the edge and connected to current by two copper electrodes at each end according to a procedure already described in previous work [2]. In experimental study, the heat transfers at the vertical surface are natural convection and thermal radiation. The temperatures at the center of the material and at the surface were measured by type K thermocouples of 250 μm diameter.

Geometry and mathematical modeling. The geometry and the mesh of the problem are show in Figure 5. The voltage is applied directly to the ends.

The material is supposed isotropic and homogenous at the macroscopic scale. The governing equations for the electrical conduction and heat transfer in the material are [7]:

$$-\nabla \cdot (\sigma(T)\nabla V) = 0 \tag{2}$$

$$d(T)C_p(T)\frac{\partial T}{\partial t} = \nabla \cdot (k(T)\nabla T) + S(T) \qquad \text{with} \quad S(T) = \sigma E^2 \tag{3}$$

with σ(T) the electrical conductivity function of temperature T; V, the voltage and S, the internal power generated by Joule effect.

The mathematical modeling is solved with the finite element method [8] using FEMLAB®.

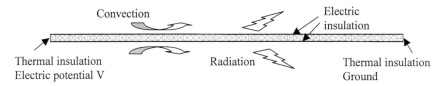

Figure 5. Longitudinal cross section, view from the top, of the extruded tape.

Confrontation of experimental data with simulating results. In order to compare experimental measurements with simulation, the voltage solicitation and air temperature with respect to time are directly read in an experimental file. Then, the thermo-physical data determined previously are brought into the model through polynomial equations. In Figure 6, the simulated intensity and temperature responses are in good agreement with experimental data. The electrical intensity was calculated from the current density J at half length of the tape. The intensity curve emphasizes the PTC effect, beginning close to 40°C and responsible for a temperature self-regulation effect. During a voltage step, the intensity instantaneously increases before it progressively decreases and reaches the steady state. It can be noticed that this phenomenon is reversed during cooling. The plotted temperatures are the experimental and simulated temperatures in the center of the tape. In this case, the heat flux is small (1 kW.m^{-2} for V = 25 V), so the thermal gradients in the material are lower than 2°C (Figure 7).

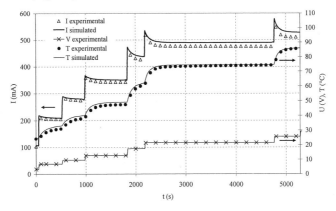

Figure 6. Comparison of experimental and simulated transient response of PP-30% w/w.

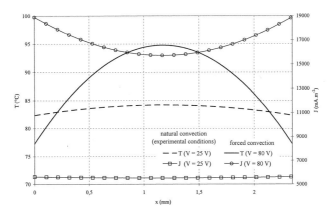

Figure 7. Current density and temperature profile through the thickness at two different voltage solicitations.

Additional information on the thermo-electrical behavior of the CPC can be deduced from Figure 7, where it appears that due to the presence of a thermal gradient between the surface and the core of the sample, the current density J is not constant in the thickness of the material. These plots obtained in stationary state conditions also show that the electrical power is generated at the surfaces (colder). The simulation validated for a low voltage (25 V) providing a low heat flux (around 1 kW.m^{-2}) allows extrapolating the behavior of the CPC at higher heat flux (9 kW.m^{-2} / 80 V). This last simulation has been made in forced convection in order to obtain 85°C in the material. In this case the temperature gradient increases from one or two degrees to about twenty so that it may be necessary to increase the thermal conductivity of the CPC.

Conclusion

PP-carbon CPC show interesting thermo-electrical properties, smooth evolution of resistivity with temperature up to 150°C and consequently high power dissipation. The addition of SCF to PP already formulated with CB increases sharply the electrical conductivity of the CPC but does not have much influence on thermal conductivity as it could have been expected from the favorable aspect ratio of the fibers. The simulations put into evidence a thermal gradient increase with the heat flux which may damage the material itself. One way to reduce this gradient is to increase the thermal conductivity. But it has been shown that thermal conductivity can only be improved by a factor 2 for a filler volume fraction of 25%. Since carbon is both electrical and thermal conductor, this method also presents the drawback to increase simultaneously electrical and thermal conductivity. Moreover, the increase of electrical filler volume fraction tends to diminish the self-regulation effect due to a better stability of conductive chains through material. This is why future developments of this work will concern the use of thermally conducting and non-electrically conducting fillers.

[1] J. F. Feller, *J Appl. Polym. Scie.*, **2004**, 91, 4, 2151.
[2] J. F. Feller, P. Chauvelon, I. Linossier, P. Glouannec, *Polymer Testing.* **2003**, 22, 7, 831.
[3] I. H. Tavman, *Int. Comm. Heat Mass Transfer.* **1998**, 25, 5, 723.
[4] J. C. Maxwell, *A treatise on Electricity & Magnetism*, Dover, New York, 1954, Chap. 9.
[5] T. Lewis, L. Nielsen, *Journal of Applied Polymer Science.* **1970**, 14, 1449.
[6] J. F. Feller., I. Linossier, S. Pimbert, G. Levesque, *Journal of Applied Polymer Science.* **2001**, 79, 5, 779.
[7] M. Necati Ozisick, *Heat Transfer: A basic Approach*, McGraw-Hill Ed, New-York, 1985.
[8] O. C. Zienkiewicz, R. L. Taylor, *Finite Element Method*, 5th ed., Butterworth-Heinemann, 2000.

Influence of the Absorbed Water on the Tensile Strength of Flax Fibers

Christophe Baley,[1] *Claudine Morvan,*[2] *Yves Grohens*[1]

[1] Université de Bretagne Sud, Laboratoire L2PIC, BP 92 116, 56 321 Lorient Cedex, France
E-mail: christophe.baley@univ-ubs.fr
[2] Université de Rouen, UMR 6037CNRS, IFRMP 23 ,-76851, Mont-Saint-Aignan Cedex, France

Summary: The complex structure of flax fibres involves many chemical biomolecules located in an amorphous matrix in which cellulose micrifibrils are embbeded. The drying of flax fibres influences significantly their tensile strength. This result can be explained by the creation of damages within the fibre and by the modification of the chemical composition of the matrix components. This loss of water involves a modification of the adhesion between the cellulose microfibrils and the matrix. This modification is due to the evolution of the components ensuring the transfer of load between the microfibrils and thus conditioning the strength of the cellular wall.

Keywords: cellulosic fibre; fibrils adhesion; tensile strength; water absorption

Introduction

Ecological concern has resulted in a renewed interest in natural materials. Cellulosic fibres such as flax, are an interesting, environmentally friendly alternative to the use of glass fibres as reinforcement in engineering composites. Flax fibres are very specialised cells, multinucleate but without septum, having extreme length (2-5cm) and thin diameter (15-25µm). When mature plant are pulled, most of the fibres are dead cells with very thick multilayer cell-walls. The most external layer consists of a primary wall (0.1-0.5 µm thick) coated with a polymer matrix (designated as middle lamellae and intercellular junctions) that insure the intercellular cohesion in a fibre bundle. The secondary wall is composed of 3 layers, S1 (0.5-2 µm thick), S2 (5-10µm) and S3 (0.5-1µm) whose main constituents are cellulose microfibrils with particular orientations [1]. Ultimate tensile strength, elastic modulus and ultimate elongation do mainly depend on both the cellulose content and the angle between microfibril and fibre axis. The mechanical properties of fibres are essentially due to the cellulose microfibrils locked into an almost axial direction (angle of 10° according to Wang[2]) in the layer S2.

The main components of primary wall and cell junctions consist of pectins, i.e. a complex polysaccharide family whose two of them principally respond to growth conditions that is 1)

© 2005 WILEY-VCH Verlag GmbH & KGaA, Weinheim   DOI: 10.1002/masy.200550425

homogalacturonans and 2) rhamnogalacturonan I (RG-I). Pectins are polyanions and hence are responsible for the main part of the adsorbed water in the cell wall, especially homogalacturonans which have a high charge density. In the cell walls, pectins organise water molecules in some kind of network of micropores and macropores [3,4]. The micropores correspond to the Donnan space where the number of water molecules depends on the nature of cations and also on the charge density while the macropores constitute a water free space where counter ions are expelled.

In contrast with the lignocellulosic fibres of wood, flax fibres almost lack lignin but cellulose microfibrils are embedded in 5-15% of non cellulosic polymers (NCPs), and are designated as cellulosic fibres. Long chains of pectic β1-4-D-galactan (DP 10-30) have been clearly identified as the main encrusting component of ultimate fibres [5-11]. They are branched on a rhamnogalacturonan backbone whose charge density is about half that of homogalacturonan. Again, these pectic components are the main components interacting with water in the secondary wall. Hemicelluloses, present in few amount, interacting strongly with the cellulose molecules at the surface of the microfibrils contribute only partly to the water adsorption. Thus, cristalline cellulose-microfibrils, coated by hemicellulosic cross-linking glucans and (galacto)glucomanans, are embedded in a pectic galactan matrix. Acidic proteins such as arabinogalactan proteins (AGPS) can also take place within the matrix of the secondary wall and contribute to water adsorption.

The interactions of water (hydrogen bonds) with fibres occur mainly via the hydroxyl groups of polysaccharides [12], at least those that are not involved in cristallites. According to Chauban [13] the adsorption of water can be divided into 2 parts : water at the surface of the fibre and ii) water within the cell walls. As described above, the level of adsoption at the fibre surface would be high and due to the large amount of pectins. It will depend on the degree of retting of flax, a microbial process that progressively degrades cell wall polysaccharides, and also of the subsequent mechanical/chemical treatments which strongly influence the surface composition [14-15]. On the other hand the adsorption of water in the secondary wall is multicomponents although essentially due to the pectic –proteoglycan matrix. Thus, the percentage of water linked to each cell wall components greatly depends on process [16] but also on variety and ecophysiology. According to the literature, the moisture of flax fibres is within a range of 6 - 10 %.

For the polymer reinforcement, the water presence in fibres is synonymous of volume variation (swelling / shrinkage), degradation, poor interfacial bonding between fibre and matrix, and water vaporisation during the process (with voids development). For the manufacture of composite material the influence of water is still an open question. In this text we are interested in the influence of a drying, therefore with a loss of water absorbed, on tensile strength of flax fibres.

Experimental

Flax plants, variety Ariane, were cultivated in Normandie (France) and retted on the field (dew retting). After harvest, the stems were scuched and the fibres combed. The percentage of absorbed water in flax fibres (n samples of x g) was calculated [as (Ww-Wd)/Wd] by weight difference before (Ww) and after drying at 105°C for 12 hours (Wd). The longitudinal tensile behaviour of flax fibres was measured using the standard method for single material (NFT 25-704, ASTM D 3379-75), by taking into account the compliance of the system. The range of the load cell is 0-2N and the precision of measure of displacement is one micron. Test conditions are: speed 1 mm/mn, and fibre length 10 mm.The tensile tests are carried out on raw fibres, drying fibres (14 hours at 105 °C) and drying fibres having reabsorbed water in contact with the ambient air (storage 8 days at 20 °C under the same conditions as raw fibres).

Results and discussion

After drying the average lost weight is 8.3%, and in contact with the ambient air the fibre reabsorbs water (Figure 1). The mass increase is rather fast and the fibres have recovered their initial mass after 5 hours.

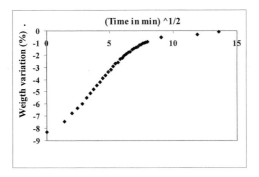

Figure 1. Evolution of the flax fibres mass in contact with the ambient air after drying (14 hours at 105°C).

198

Mechanical properties of raw flax fibres were presented in a preceding article [17], we notice that Young's modulus and tensile strength are a function of fibre diameter. The tensile strength are indicated for two classes of fibres diameter (20-22.5 μm and 22.5-25 μm) according to the treatments undergone (Figures 2 and 3). One notices, for the two classes of fibres, the influence of drying on tensile strength which results in a the fall of the mechanical properties. After water absorption, the fibres do not recover their initial properties. This result can be explained by the creation of damage within fibre and by modification of the chemical components location in the fibres. In addition the results are more dispersed than before drying. Dispersions of fibre properties are due to two main factors: 1) the fibre development in the stem and 2) the transformation process.

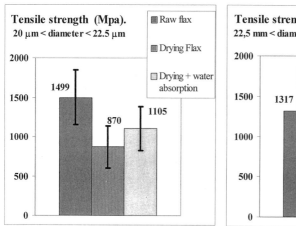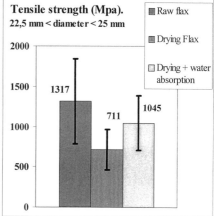

Figures 2 and 3. Drying influence on tensile strength of flax fibres.

To illustrate the influence of the absorbed water by a flax fibre on the damage mechanisms, a node is made on fibre, then observed under SEM. The operation is carried out on a raw fibre and on a fibre after drying (14 hours with 105 °C). For a fibre with moisture content of 8.3 % (raw fibre), we notice two damage mechanisms (Figures 4 and 5): buckling in the compression area and cracking in the tensile area (these cracks are explained simply by the Poisson's ratio of fibre). After drying, we notice a loss of cohesion between the matrix and the microfibrils which seems to be mainly influenced by the absorbed water (Figures 6 and 7).

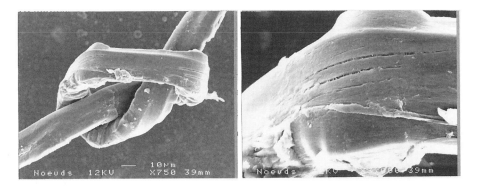

Figures 4 and 5. Damages in a flax fibre with a moisture percentage of 8.3 %.

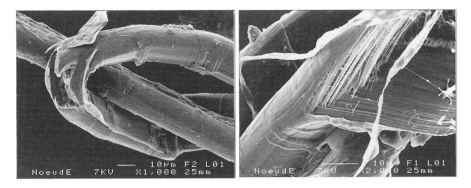

Figures 6 and 7. Damages in a flax fibre after drying.

Astley and Donald [18] studied the effect of the hydration on the distribution of cellulose microfibrils of the S2 layer (secondary wall) of a flax fibre using small angles x-rays scattering. Water penetrates in the amorphous zones and enters in competition for the sites of potential hydrogen bonds, involving a loss of the interactions between crystalline areas (cellulose microfibrils) and the amorphous matrix. This effect involves an increase in the strain at failure as reported by [19] and Joffe [20] who considers that the absorption of water has a plasticizing effect on a flax fibre In addition, Netravali [21] claimed, without indicating experimental results, that repeated moisture absorption/desorption cycles can significantly lower the fibres strength. The percentage of water absorptive in mass by a flax fibre is general information because this percentage is not homogeneously distributed in the thickness of fibre, the exchanges with the air being done on the surface [22]. Moreover, its structure is microporous which influences the retention of water [23].

The thermal stability of flax fibres is a significant parameter for the processing of composite materials and it is generally admitted that the low thermal stability of vegetable natural fibres presents a limit at their use [24]. This is particularly true for the thermoplastic matrices whose process temperatures are high. Therefore, during the temperature increase one notes:

- a vaporisation of absorbed water,
- the development of internal mechanical strains. The thermal expansion coefficients of the components of the cellular walls are different. An increase in temperature will create irreversible damages and cracks affecting the physical properties of the fibres [25],
- Physicochemical modifications of the components, and possibly a degradation beyond a certain temperature.The loss of water starts at 60°C and at 120°C there is no more water and there is a degradation of waxes, at 180°C there is a decomposition of pectins and at 230°C a degradation of hemicelluloses and cellulose is claimed [26].

The definition of the acceptable temperatures of transformation of vegetable fibres is studied by several authors. The evolution of the fibres properties is a function of the temperature and the time. 200°C mustn't be exceed for more than 5 min (current data) for vegetable fibres [20,27] and the ideal is to not exceed 160°C which is a low temperature for thermoplastic processing.

Conclusions

The drying of flax fibres influence significantly the tensile strength and after water desorption/absorption cycle the fibres do not recover their initial properties. This result can be explained by the creation of damages within the fibre and by the modification of the chemical composition of the matrix components. This loss of water involves a modification of cohesion between the cellulose microfibrils, modification due to the evolution of the components ensuring the transfer of load between the microfibrils and thus conditioning the strength of the cellular wall. For the processing of composite materials, the limited thermal stability of flax fibres is a key parameter to be taken into account for the choice of polymer and the thermal cycle. It is worth noting that such mild and specific treatments could keep to fibres the maximum of their native properties.

[1] J.C. Roland, M. Mosiniak, D. Roland D, *Acta Bot. Gallica*, **1995**, 142, 463.

[2] H.H. Wang, J.G. Drummont, S.M. Reath, K. Hunt, P.A. Watson, *Wood Science and Technology*, **2001**, 34, 493.

[3] J. Dainty, A.B. Hope, *Aust J Biol Sci*, **1961**, 14, 541.

[4] M. Demarty, C. Morvan, M. Thellier, *Plant Cell environment*,**1984**, 7, 441.

[5] C. Morvan, C. Andème-Onzighi, R. Girault, D.S. Himmelsbach, A. Driouich, D.E. Akin, *Plant Physiol. Biochem.*, **2003**, 41, 935.

[6] C. Andème-Onzighi, R. Girault, I. His, C. Morvan, A. Driouich, *Protoplasma*, **2000**, 213, 235.

[7] R. Girault, I. His, C. Andème-Onzighi, A. Dirouich, C. Morvan, *Planta*, **2000**, 211, 256.

[8] T.A. Gorshkova, S.B. Chemikosova, V.V. Lozovaya, N.C. Carpita, *Plant Physiol.*, **1997**, 114, 723.

[9] J.M. Van Hazendonk, E.J.M. Reinerink, P. de Waard, J.E.G. van Dam, *Carbohydr. Res.*, **1996**, 291, 141.

[10] I. His, C. Andème-Onzighi, C. Morvan, A. Driouich, *J. Histochem. Cytochem.*, **2001**, 49, 1525.

[11] C. Mooney, T. Stolle-Smits, H. Schols, E. de Jong, *J. Biotech.*, **2001**, 89, 205.

[12] S. Das, A.K. Saha, P.K. Choudhury, R.K. Basak, B.C. Mitra, T. Todd, S. Lang, R.H. Rowell, *Journal of Applied Polymer Science*, **2000**, 76, 1652.

[13] S.S. Chauban, P. Aggarwal, A. Karmarkar, K.K. Pandey, *Holz ah Roh-und Werkstoff*, **2001**, 59, 250.

[14] C. Morvan, A. Abdul Hafez, O. Morvan, A. Jauneau, M. Demarty,. *Plant Physiol. Biochem.*, **1989**, 27, 451.

[15] A. Jauneau, F. Bert, C. Rihouey, C. Morvan, *Biofutur*, **1997**, 167, 34.

[16] F. Thuvander, G. Kifetew, L.A. Berglund, *Wood Science and Technology*, **2002**, 36, 241.

[17] C. Baley, *Composites Part A*, **2002**, 33, 939.

[18] O.M. Astley, A.M. Donald, *Biomacromolecules*, **2001**, 2, 672.

[19] L. Köhler. Natural cellulose fibers : properties. In. *"Encyclopedia of materials ; Science and Technology"* Eds., Elsevier Science Ltd, **2001**, p. 5944.

[20] R. Joffe, J. Andersons, L. Wallström, *Composites Part A*, **2003**, 34, 603.

[21] A. Netravali, S. Chabba, *Materials Today*, April **2003**, 22.

[22] S. Pang, *Wood Science and Technology*, **2002**, 36, 75.

[23] K.M. Mannan, M.A.I. Talukdar, *Polymer*, **1997**, 38, 10, 2493.

[24] H. Lilholt, J.M. Lawther. Natural Organic Fibers. in: *"Comprehensive Composite Materials"*, Pergamon, Edts., Elsevier Science, **2000**, Chap. 1-10.

[25] J. Gassan, A. Chate, J.K. Bledzki, *Journal of Materials Science*, **2001**, 36, 3715.

[26] K. Van de Velde, E. Baetens, *Macromelcular Materials and Engineering*, **2001**, 286, 342.

[27] S.C. Jana, A. Prieto, *Journal of Applied Polymer Science*, **2002**, 86, 2168.

Macromol. Symp. **2005**, *222*, 203-208

Stabilisation Effect of Calcium Ions on Polymer Network in Hydrogels Derived from a Lyotropic Phase of Hydroxypropylcellulose

Aleksandra Joachimiak,* Lidia Okrasa, Tomasz Halamus, Piotr Wojciechowski

Department of Molecular Physics, Technical University of Lodz Zeromskiego 116, 90-924, Lodz, Poland
Fax: + 48 42 631 32 18, E-mail: ajoachim@p.lodz.pl

Summary: The process of crosslinking of the hydrogel, derived from lyotropic liquid crystalline (LLC) phases of hydroxypropylcellulose/acrylic acid-water, by calcium ions was studied by means of Raman and dielectric spectroscopy. Formation of salt by poly(acrylic acid) and calcium ions, resulting in hydrogel crosslinking, induces differences in Raman spectra of the hydrogel before and after the crosslinking. The crosslinking results in significant increase in the activation energy of β-relaxation of poly(acrylic acid). This is a direct consequence of restriction in motions of carboxylic groups of poly(acrylic acid) due to calcium salt formation. Thus, the crosslinking improves polymer network stability in the hydrogel in the swollen state.

Keywords: dielectric spectroscopy; hydrogels; hydroxypropylcellulose; Raman spectroscopy

Introduction

Hydrogels, three-dimensional polymer networks, are able to absorb a large amount of water, which makes them interesting and promising materials offering various application possibilities, especially in medicine, pharmacy, and biotechnology.[1,2] However, the application possibility depends on the swelling properties and stability of the polymer network in the swollen state in defined environment conditions. The low density of chemical crosslinks very often causes poor stability of the polymer network of the hydrogel in the swollen state. Therefore, long immersion processes of such hydrogels can lead to destruction of the polymer network.

Recently, special emphasis has been put on hydrogels with liquid crystalline organisation of polymer network because of their similarity to the gel networks present in living organisms.[3] As it was reported in our previous paper[4], the photopolymerisation of acrylic acid in the lyotropic liquid crystalline (LLC) phase of hydroxypropylcellulose in the mixture of acrylic acid (AA) and water leads to a formation of optically anisotropic hydrogel. The process of photopolymerisation of AA in lyotropic phase is faster than phase separation and, therefore,

 DOI: 10.1002/masy.200550426

the hydrogel sample is transparent, although both polymers - HPC and poly(AA) - are not miscible. However, the obtained hydrogel exhibits limited stability of the polymer network during long-term immersion in the water. This is an outcome of the fact that the network of HPC-poly(AA)-H_2O hydrogel is stabilised only by hydrogen bonds between HPC, poly(acrylic acid) and water, which is in agreement with the results obtained previously for the anisotropic composite of HPC-poly(AA) derived from photopolymerisable lyotropic phase of HPC/AA.[5]

In this paper, the process of crosslinking of the hydrogel – derived from lyotropic phase of HPC/AA-H_2O by calcium ions – and its impact on the stability of the polymer network in the swollen state is presented. We have taken advantage of poly(AA) ability to creat a salt with metal ions[6], such as calcium. The influence of crosslinking of the HPC-poly(AA)-H_2O hydrogel by calcium ions on the relaxation phenomena of the polymer network is also discussed.

Experimental Part

(2-hydroxypropyl)cellulose (M_w=100 000 g/mol), acrylic acid and calcium acetate were supplied by Aldrich Chemical Co. Acrylic acid was further purified by distillation.

The hydrogel HPC-poly(AA)-H_2O was the object of the research. This compound was prepared by photopolymerisation (λ=365 nm) of AA in the LLC-phase of HPC/AA-H_2O (40:40:20 wt.%), according to the procedure described elsewhere.[4] The photoinitiator: Esacure 651 (Ciba-Geigy) was added in the amount of 0.3 wt.% on the basis of the AA weight. The hydrogel, obtained in such a way, was afterwards crosslinked by calcium ions (Ca^{2+}), by means of immersion of the hydrogel in the water solution of calcium acetate with Ca^{2+} concentration of 0.22 mol×dm^{-3}, during 72 hours at the room temperature. After immersion, the hydrogel samples were dried at the room temperature.

The Raman analysis was performed using dispersive spectrometer Jobin-Yvon T64000, equipped with confocal microscope, at 293 K using Argon laser wavelength of λ=514 nm.

Dielectric relaxation spectroscopy was performed in Max Planck Institute for Polymer Research in Mainz (Germany) by means of Alfa High Resolution Dielectric Analyzer (Novocontrol GmbH), for the temperature range of 130-430 K and the frequency range of 10^{-1}-10^6 Hz.

The swelling study of the hydrogel (diameter $2R$=20 mm, thickness d=0.2 mm) was carried out at the room temperature by registering changes of water uptake (h=immersed water weight/dry hydrogel weight) vs. time.

Results

Figure 1a shows the effect of immersion time on the water uptake, obtained for the hydrogel crosslinked by calcium ions. This dependence is compared with the results reported earlier for the non-crosslinked hydrogel (HPC-poly(AA)-H_2O).[4] The hydrogel crosslinked by calcium ions has approximately constant value of the water uptake (h) even after longer than one month immersion in water, which indicates better stability of the polymer network of the crosslinked hydrogel in the swollen state, with respect to the non-crosslinked hydrogel. However, three-dimensional polymer network of the non-crosslinked hydrogel is destroyed after long immersion, which is a direct consequence of destruction of hydrogen bonds between HPC and poly(AA).

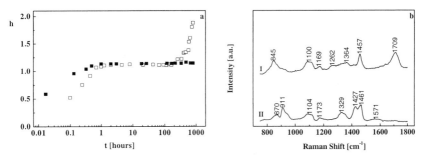

Figure 1. a) Dependence of water uptake on time for the hydrogels: non-crosslinked (open symbols) and crosslinked (solid symbols) by calcium ions. b) The Raman spectra of the dry sample of the hydrogel, non-crosslinked (line I) and crosslinked by calcium ions (line II).

The process of formation of calcium salt of poly(AA) was detected by means of Raman spectroscopy, when comparing the Raman spectra of the hydrogel before and after crosslinking by calcium ions (Figure 1b). The band at 1709 cm^{-1}, related to the carbonyl group of poly(AA), seen in the Raman spectrum of the non-crosslinked hydrogel (line I) disappears after crosslinking by calcium ions. At the same time, a new band at 1427 cm^{-1} comes to existence (line II). This band is typical for calcium salt of poly(AA) and is related to symmetric vibration of -COO⁻ group (carboxylate anion) of created salt. Our results are similar to the Raman spectra of poly(AA) and its salts with calcium, sodium and aluminium at the range of 1400-1800 cm^{-1}, reported by M. Young et al.[6] The Raman spectroscopy results provide the evidence for the calcium salt formation by poly(AA) present in the HPC-poly(AA)-H$_2$O hydrogel.

It should be emphasised that the process of crosslinking of the hydrogel by calcium ions does not have a perceptible impact on mesomorphic organisation of the polymer network of the hydrogel. Therefore, this problem will not be discussed in this paper.

It is believed that formation of chemical bonds between poly(AA) and calcium ions in the hydrogel network affects the molecular motions of the hydrogel network, especially poly(AA). The relaxation processes of the polymer network in the hydrogel, before and after crosslinking, were analysed by dielectric relaxation spectroscopy.

Figure 2a shows the dependence of real (M') and imaginary (M'') parts of electric modulus on temperature for the non-crosslinked hydrogel, containing water originating from LLC-phase of HPC/AA-H$_2$O as well as after its drying. Additionally, Figure 2b presents the activation map of relaxation times of the hydrogel.

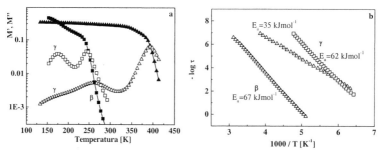

Figure 2. a) Temperature dependence of M' (solid symbols) and M'' (open symbols) at f=1 kHz, and b) activation map obtained for the hydrogel containing 20 wt.-% of water from LLC-phase (squares) and after drying (triangles). $\tau = 1/(2\pi f)$, where τ - relaxation time and f – frequency.

In case of the hydrogel containing water coming from LLC-phase, two bands in the temperature dependence of M' and M'' are observed. The band at 175 K (at 1 kHz) can be considered, according to the literature[7], as γ-relaxation of HPC, related to the local motions of side chains (hydroxypropyl groups) of HPC. The activation energy of registered γ-relaxation of HPC of the anisotropic hydrogel equals E_a=62 kJ×mol^{-1}. This value is higher than the activation energy of γ-relaxation of pure HPC, reported by M. Pizzola (E_a=41 kJ×mol^{-1}).[7] The difference between the obtained and literature value of E_a of γ-relaxation of HPC is a consequence of the presence of water molecules in the HPC-poly(AA)-H$_2$O hydrogel network. The hydration shell created by water molecules arround hydroxypropyl groups of HPC in the hydrogel network by results in significant increase in the E_a value of γ-relaxation of HPC. The band, observed at the temperature of 245 K (at 1 kHz), was not considered as it relates to the ionic conductivity. However, it should be emphasised that the amount of water in the hydrogel significantly influences the ionic conductivity.

In case of the dried hydrogel, the ionic conductivity phenomenon appears at much higher temperatures equal to 400 K (Figure 2a), which results from limited amount of water molecules in the hydrogel. Apart from the ionic conductivity effect, seen at 1 kHz at 400 K, a band at 255 K in the temperature dependence of M' and M'' is also observed (Figure 2a). Dielectric spectroscopy results of the HPC-poly(AA) composites, reported by L. Okrasa et al.[5], indicate this band as the one related to β-relaxation of poly(AA). This relaxation process relates to the local motions of carboxylic groups of poly(AA)[5]. The value of E_a of β-relaxation, assessed for the dry hydrogel, equals E_a=67 kJ×mol^{-1} and is similar to the literature value (E_a=56 kJ×mol^{-1}).[5] The band of γ-relaxation of HPC in the dry hydrogel network is seen at 180 K in the temperature dependence of electric modulus at 1 kHz (Figure 2a). The assessed E_a of γ-relaxation of HPC of dry hydrogel network (E_a=35 kJ×mol^{-1}) does not differ significantly from the value reported by M. Pizzola (E_a=41 kJ×mol^{-1}).[7] One can conclude that the presence of water molecules in the hydrogel influences not only the ionic conductivity but also the γ-relaxation of HPC.

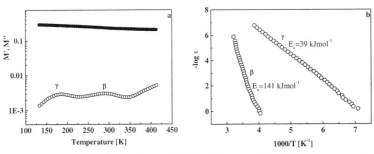

Figure 3. a) Temperature dependence of M' (solid symbols) and M'' (open symbols) at f=1 kHz, and b) activation map obtained for dried hydrogel crosslinked by Ca^{2+}.

Figure 3a presents the influence of the temperature on the electric modulus of the hydrogel after its crosslinking by calcium ions and drying. Additionally, Figure 3b presents the activation map, showing relaxation times of this hydrogel network. For the crosslinked hydrogel, two bands: β-relaxation of poly(AA) and γ-relaxation of HPC are seen in the temperature dependence of electric modulus, similarly as it was observed for non-crosslinked hydrogel network. However, β-relaxation of poly(AA) shifts slightly at 1 kHz to 287 K after crosslinking process. Additionally, after crosslinking of the anisotropic hydrogel by calcium ions, the significant change in the value of activation energy of β-relaxation of poly(AA) is observed (Figure 3b). Due to crosslinking of the anisotropic hydrogel by Ca^{2+} ions, E_a of β-relaxation of poly(AA) (67 kJ×mol^{-1} for non-crosslinked hydrogel) increases to the value of 141 kJ×mol^{-1}. This suggests that the motions of -COOH group of poly(AA) in the polymer network of crosslinked hydrogel are strongly restricted, which can result from the process of formation of calcium salt of poly(AA), well revealed by Raman spectroscopy (Figure 1b). The value of E_a of γ-relaxation of HPC does not increase significantly (from 35 for non-crosslinked to 39 kJ×mol^{-1} for crosslinked network) which suggests that calcium ions interact chemically during crosslinking process mostly with poly(AA).

Conclusion

Stabilisation of the polymer network of the HPC-poly(AA)-H_2O hydrogel in the swollen state results from the formation of chemical bonds between Ca^{2+} and poly(AA) present in the hydrogel, which has been revealed by Raman spectroscopy analysis. Hydrogel crosslinking leads to an increase in activation energy of relaxation processes of the main polymer constituents of the hydrogel network. This is especially visible for β-relaxation of poly(AA), present in the hydrogel network, which is a consequence of restriction in motions of polyacid carboxylic groups. Additionally, the water present in the hydrogel causes significant increase in the activation energy of γ-relaxation of HPC, related to the motions of side chains of polysaccharide chain of HPC, as well as shifts the ionic conductivity effect towards lower temperatures.

[1]. J. M. Rosiak, P. Ulański, L. A. Pajewski, F. Yoshii, K. Makuuchi, *Radiation Physics and Chemistry*, **1995**, *46*, 161.
[2]. I. V. Galev and B. Mattiasson, *Trends in Biotechnology.*, **1999**, *17*, 335.
[3]. T. Kaneko, K. Yamaoka, J. P. Gong, Y. Osada, *Macromolecules*, **2000**, *33*, 412.
[4]. P. Wojciechowski, A. Joachimiak, T. Halamus, *Polymers For Advanced Technologies*, **2003**, *14*, 826.
[5]. L. Okrasa, G. Boiteux, J. Ulanski, G. Seytre, *Polymer*, **2001**, *42*, 3817.
[6]. A. M. Young, A. Sherpa, G. Pearson, B. Schottlander, D. N. Waters, *Biomaterials*, **2000**, *21*, 1971.
[7]. M. Pizzoli, M. Scandola, G. Ceccorulli, *Plastics, Rubber and Composites Processing and Applications*, **1991**, *16*, 239.

Macromol. Symp. **2005**, *222*, 209-217

Peculiarities of Formation of Intermolecular Polycomplexes Based on Polyacrylamide, Poly(vinyl alcohol) and Poly(ethylene oxide)

Lesya Momot, *Tatyana Zheltonozhskaya, Nataliya Permyakova, Sergey Fedorchuk, Vladimir Syromyatnikov*

Macromolecular Chemistry Department, Kiev Taras Shevchenko National University, 64 Vladimirskaya Str., 01033, Kiev, Ukraine
E-mail: momot_lesya@ukr.net

Summary: The phenomenon of self-assembly of aggregates formed by relatively short chains of poly(vinyl alcohol) (PVA) on the long macromolecules of polyacrylamide (PAA) in aqueous medium are discussed. PVA and PAA form intermolecular polycomplexes (InterPC) of a constant composition independently on a ratio of polymer components. The complex formation between high-molecular-weight PAA and relatively low-molecular-weight poly(ethylene oxide) (PEO) are considered also. PEO with $M \leq 4 \cdot 10^4$ g.mol^{-1} weakly interacts with PAA. The polymer-polymer interaction can be intensified when the part of amide groups (~20 mol %) on PAA chain to transform into the carboxylic groups. InterPCs formed by PEO and initial or modified PAA have associative structure with friable packing of the polymer segments. They are stabilized by the hydrogen bond system.

Keywords: aggregates; hydrogen bonds; intermolecular polycomplexes; matrices; polyacrylamide; poly(ethylene oxide); poly(vinyl alcohol)

Introduction

In spite of numerous successes achieved at establishment of the main conformities of polymer-polymer interactions, researches of the intermolecular polycomplex (InterPC) structures connected with their properties remain an actual scientific problem. Polymer-polymer complexes stabilized mainly by the electrostatic interactions (the polyelectrolyte InterPC), today are widely studied.[1] InterPCs formed first of all by the hydrogen bonds are lesser known.[2,3] But such polycomplexes are used in designing of new medicines, as separating membranes, flocculants and other important functional materials.[4]

Processes of polymer-polymer interactions between very long (matrices) and relatively short ("oligomers") macromolecules are of special interest because they allow to find so called "the critical polymer chain length" which is necessary to achieve for the cooperative interaction between two polymers:[5]

$$P_1 \text{ (matrice)} + n \cdot P_2 \rightarrow \text{InterPC } (P_1 + n \cdot P_2)$$

In the present work complex formation in two pairs of hydrophilic polymers such as polyacrylamide (PAA) - poly(vinyl alcohol) (PVA) and PAA - poly(ethylene oxide) (PEO) are considered.

Reactions of formation and structure of InterPCs

The basic parameters of PVA, PAA and PEO samples under study are shown in Table 1-2.

Table 1. Molecular parameters

System of PAA+PVA			
Polymer	$M_v \cdot 10^{-6}$	$a^{1)}$ %	$b^{2)}$ %
PVA 1	0.04	-	33
PVA 2	0.08	-	13
PVA 3	0.12	-	31
PAA 1	2.72	11	-
PAA 2	4.40	1	-

[1)]The degree of hydrolysis of acrylamide links.

[2)]The quantity of residual acetate groups in PVA.

Table 2. Molecular parameters

System of PAA+PEO		
Polymer	$M_v \cdot 10^{-5}$	$a^{1)}$ %
PAA3	9.80	1
P(AAm-co-AAc)	11.4	20
PEO 1	0.02	-
PEO 2	0.04	-
PEO 3	0.06	-
PEO 4	0.4	-
PEO 5	1.49	-

[1)]The degree of hydrolysis of acrylamide links.

It is seen that PAA length surpasses essentially the length of PVA or PEO chains. All the samples of PAA were synthesized by the acrylamide ("Merck", Germany) radical polymerization with using Ce^{IV} salt as initiator. The alkaline hydrolysis of PAA was carried out to obtain P(AAm-co-AAc) the random acrylamide with acrylic acid copolymer having the hydrolysis degree ~20%. The samples of PVA (Japan) and PEO ("Merck", Germany) were used also. The complex formation reactions were carried out by mixing of PAA with PVA or PEO in aqueous solutions during 1 hour.

System of PAA+PVA. The complex formation between PAA and PVA is displayed first of all in the viscometry data (Figure 1). Viscosity of polymer solutions was measured by Ostwald-type viscometer (τ_0=94 s at T=298 K). The deviation of the value $\eta_{sp\ mix}/\Sigma\eta_{sp\ i}$ from unity in a wide region of the mixture compositions testifies about the interaction of components. Positive deviations (curves 1, 2) point on to formation of InterPC particles with friable packing of the component segments, but negative ones (curves 3, 4) testifies to arising of the compact complex particles. The ratio φ between the component concentrations in

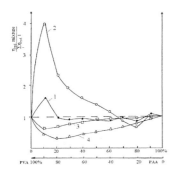

Figure 1. The ratio of η_{sp}mixture/$\sum\eta_{sp}i$ *vs* mixture composition for PAA2+PVA1 – *1*, PAA1+PVA1 -*2*, PAA1+PVA2 -*3* and PAA1+PVA3 -*4*. T=298 K.

extreme points of the curves can indicate on φ_{char} every InterPC, at which both polymers are connected quantitavely with each other.

This assumption was confirmed earlier[6] for the compact particles in the PAA2+PVA1 mixture by gel chromatography.

In the present study validity of such assumption for the friable InterPC particles in the PAA1+PVA2 and PAA1+PVA3 mixtures by the high-speed sedimentation data (Figure 2,3; Table 3,4) was proved.

The sedimentation coefficients of individual polymers and their mixtures were determined with the use of analytical centrifuge MOM 3170 H (Hungary).

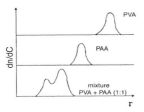

Figure 2. Sedimentograms of PVA2, PAA1 and their mixture (1:1) after 45 min centrifugation. ω=4·10^4 rs·min^{-1}, T=295 K, C_{PVA}=C_{PAA}=0,5 kg·m^{-3}.

Examples of sedimentograms are shown in Figure 2. Sedimentograms of the mixtures contained two peaks excluding the 9:1 ratio, which corresponds to φ_{char} (Table 4). The excess of bound PAA (matrice) for all the ratios φ= C_{PVA}/C_{PAA}<φ_{char} was displayed as a separate sedimentation peak (Figure 2). The values of $1/S_c$ and Q (the area of the sedimentation peak) for the individual

polymers linearly grow with concentration (Figure 3 a, Table 4), that is characteristic for the dense macrocoils, which do not change their state in viewed concentration field.[7] Q values for the 1-st peaks surpass considerably that ones for each component (Table 3) that allows to consider their as peaks of InterPC. The areas of the 2-nd peaks in both mixtures were lesser than peaks of separate components. The linear concentration dependences $1/S_c$ and Q for the 1-st peaks (Figure 3 b, lines 1,2; Table 4) were observed at all the mixture compositions. This fact allows concluding that the composition of both InterPCs in surveyed regions of concentrations and polymer ratios are constant. These peaks belong to unbound PAA. The linear dependence of $1/S_c$ for the 2-nd peaks from C_{PAA}, which are surplus concentration of PAA comparing with PAA concentration in corresponding InterPC (Figure 3 b, lines 3,4)

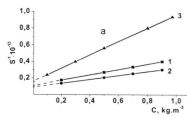 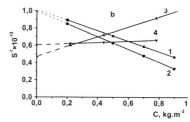

Figure 3. The inverse value of the sedimentation coefficient *vs* concentration for PVA2 –*1* (a), PVA3 –*2* (a), PAA1 –*3* (a) and mixtures of PVA2+PAA1 (1-st peak) –*1* (b), PVA3+PAA1 (1-st peak) - *2* (b), PVA2+PAA1 (2-nd peak) –*3* (b) and PVA3+PAA1 (2-nd peak) –*4* (b). T = 295 K.

confirms this conclusion. It is necessary to mark a negative declination of the line $1/S_c=f(C)$ for both InterPC (Figure 3 b, lines 1,2). Accordingly to Ref.[7], this effect is one more certificate to friable packing of polymer segments in formed InterPC. The comparison of S_0 values in Table 3 shows that complex formation of PAA1 with PVA2 and PVA3 is accompanied by destroying of their aggregates in aqueous solution. It is seen also that low-molecular-weight PVA is aggregated in initial state much more than high-molecular-weight PAA.

Table 3. Sedimentation coefficients for separate polymers and polymer mixtures

Composition	PVA 2	PVA 3	PAA 1	$S_c^{1)} \times 10^{13}$ s			
				PAA1+PVA2		PAA1+PVA3	
w_{PVA}/w_{PAA}				1st peak	2nd peak	1st peak	2nd peak
9:1	2.50	3.32	4.22	2.14	-	3.01	-
7:3	3.19	3.93	2.51	1.70	1.66	2.08	1.61
5:5	3.84	4.82	1.78	1.41	1.39	1.59	1.57
2:8	5.70	7.32	1.25	1.12	1.09	1.18	1.51
1:30	-	-	1.06	-	0.97	-	-
$S_0^{2)} \times 10^{13}$ s	9.39	11.11	6.42	0.99	2.12	1.00	1.65

[1)] The concentration coefficient of sedimentation.
[2)] The sedimentation coefficient, extrapolated to C=0.

Table 4. The areas of the sedimentation peaks for polymers and polymer mixtures

Composition	PVA2	PVA 3	PAA1	Q in conditional units	
				PAA1+PVA2	PAA1+PVA3
w_{PVA}/w_{PAA}				1st peak	2nd peak
9:1	57	44	6	67	68
7:3	44	34	20	64	59
5:5	32	24	28	47	36
2:8	13	9	47	21	15

The values of φ_{char} found in such a way are shown in Table 5.

Table 5. The characteristic compositions of InterPCs

InterPC	φ_{char}			φ'_{char} [1]
	$w_{PVA}/$ w_{PAA}	$mole_{PVA}/$ $mole_{PAA}$	base-$mole_{PVA}/$ base-$mole_{PAA}$	$mole_{PVA}/$ $mole_{PAA}$
PAA1+PVA1	9	611	12	50
PAA1+PVA2	9	306	14	22
PAA1+PVA3	9	204	12	17
PAA2+PVA1	4	444	5	82

[1] Calculated compositions of InterPC in the assumption of the full filling of PAA chain (a matrice) by the stretched PVA chains (1:1 base-$mole_{PVA}/$ base-$mole_{PAA}$).

It is seen that M_{vPVA} does not influence on φ_{char} of InterPC but it changes the packing density of the polymer segments. At the same time φ_{char} decreases and the packing density of segments in InterPC particles increases when M_{vPAA} grows. Characteristically that all the InterPCs studied contain a considerable excess of PVA. This fact and also the effect of destroying of the PVA aggregates during complex formation points on interaction of the long PAA macromolecules (matrices) with partially diminished aggregates of PVA. On the other hand, every InterPC have the constant φ_{char} value independently from the mixture composition. This results in appearance of the unbound PAA macromolecules in the range $\varphi < \varphi_{char}$. Hence, the diminished PVA aggregates are unevenly distributed among the PAA of matrices in the process of complex formation, that is the phenomenon of self-assembly of PVA aggregates on the matrices is observed.

It has been established by IR spectroscopy[6] that a major factor of formation of InterPC particles is the system of intermolecular hydrogen bonds such as:

Moreover, the hydrophobic interactions act as additional factor of stabilization of the InterPC compact particles. Really, it has been established by the benzene solubilization that such particles contain a developed hydrophobic regions.[6]

System of PAA+PEO. According to the viscometry data PEO with relatively short polymer chains ($M \leq 4 \cdot 10^4$) weakly interacts with PAA, that is why the specific viscosity of PAA solutions insignificantly increases at the PEO addition (Figure 4, curves 1,2). The appreciable growth of specific viscosity of PAA solutions at PEO addition begins at $M_{vPEO} > 1 \cdot 10^5$

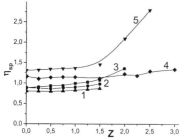

Figure 4. Specific viscosity *vs* mixture composition: for PAA3+PEO3 -*1*, PAA3+PEO4 -*2*, PAA3+PEO5 -*3* (C_{PAA3}=1 kg·m^{-3}) and P(AAm-*co*-AAc) +PEO3 -*4*, P(AAm-*co*-AAc)+PEO5 –*5* ($C_{P(AAm-co-AAc)}$=0.2 kg·m^{-3}). T=298 K.

(Figure 4, curve 3). It means that the critical M_{vPEO} ("critical PEO length") in the reaction with PAA is about $1·10^5$. Such value large comparing with the critical M_{vPEO} value in reactions with poly(acrylic acid) and poly(methacrylic acid)[8] that is conditioned by more weak interaction of PEO with PAA.

In order to intensify the connection of PAA with PEO some part (~20 mol %) of the PAA amides groups were transformed into carboxylic groups by the reaction of alkaline hydrolysis, which was carried out at T=323 K during 4 hours (concentration of PAA C=10 kg·m^{-3}). More strong binding of modified PAA with PEO is reflected by the sharp growth of the η_{sp} *vs* Z (the ratio between base-mol PEO and base-mol initial or modified PAA in the mixture) in Figure 4, curve 5. Note that InterPC particles formed by the initial or modified PAA and PEO are characterized by friable packing of polymer segment.

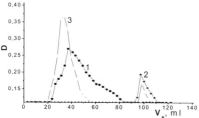

Figure 5. Gel chromatograms of aqueous solutions of P(AAm-*co*-AAc) -*1*, PEO3 -*2* and their mixture -*3* at Z=1.5; λ=220 nm, T=298 K.

Complex formation between P(AAm-*co*-AAc) and PEO is confirmed also by gel chromatography data (Figure 5). It is seen, that the elution peak of P(AAm-*co*-AAc) shifts to lower V_e values in the polymer mixture (curve 3). Simultaneously, the intensity of the PEO elution peak essentially reduces.

Existence of the H-bond system between P(AAm-*co*-AAc) and PEO was established by IR spectroscopy. The IR spectra of thin polymer films of PAA3 and P(AAm-*co*-AAc) and also the mixture P(AAm-*co*-AAc)+PEO5 on fluorite glasses (l=7-9 μm) were recorded on "Micolet NIXUS-475" spectrometer (USA) in the range 1000-4000 cm^{-1} (two examples in Figure 6). The separation of the strongly overlapped vibration bands in the Amide I, Amide II region was carried out by the spline method (Figure 7).[9] Note, that correct separation of the complicate band contour in this region was impossible taking into account only three most intense bands. That is why computer analysis of this region based on one wide band of the Amide II, two visible bands of $\nu_{C=O}$ vibration of COOH groups (at $\nu > 1700$ cm⁻¹) and three well known bands of the Amide I corresponding to *cis-trans-* and *trans*-multimers of amide groups.[10] Results of calculations ara presented in Table 6.

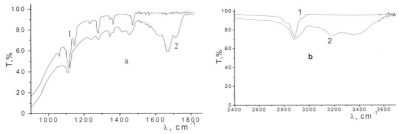

Figure 6. IR spectra of PEO5–*1* and the mixture P(AAm-*co*-AAc)+PEO5 –*2* in the regions of Amide I, Amide II and and ν_{C-O-C} vibrations (a) and also ν_{C-H}, ν_{N-H} and ν_{O-H} vibrations (b).

Table 6. The contributions (α) of separate $\nu_{C=O}$ bands of individual polymers and polymer mixture in common $\nu_{C=O}$ absorption

Sample	α %						$\beta^{1)}$
	$\nu \sim 1662$	$\nu \sim 1678$	$\nu \sim 1690$	$\nu \sim 1708$	$\nu \sim 1711$	$\nu \sim 1723$	
	cm⁻¹	cm⁻¹	cm⁻¹	cm⁻¹	cm⁻¹	cm⁻¹	
PAA3	66.9	10.7	21.1	1.3	-	0.1	0.50
P(AAm-*co*-AAc)	35.7	6.4	7.5	-	25.0	5.1	0.85
P(AAm-*co*-AAc)+ PEO5	38.8	10.4	5.0	-	20.9	24.9	2.10

1) Effective length of *trans*-multimers of amide groups. $\beta = B_{1678}/B_{1690}$, where B_i is the apparent integral absorption coefficients.

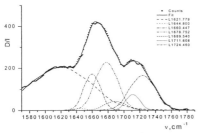

Figure 7. The example of computer processing of IR spectrum of P(AAm-*co*-AAc)+PEO5) mixture in the Amide I and Amide II region. Experimental (····) and calculated (–) vibration band contours.

The initial PAA3 as other PAA sample contained: i) the most contribution α from *cis-trans*-multimers of amide groups (the band of 1662 cm⁻¹), ii) lesser contribution from *trans*-multimers of amide groups (two bands at 1678 and 1690 cm⁻¹), iii) small contribution of free amide groups (the band of 1708 cm⁻¹) and iiii) very low contribution of COOH groups forming the "open dimer" structure of H-bonds. In the P(AAm-*co*-AAc) sample the quantities of *cis-trans-* and *trans*-multimers of amide groups sharply decreased (Table 6), but that one of COOH groups grew. These groups formed not only "open dimers", but also the mixed cyclic dimmers with amide groups (the band of 1711-1712 cm^{-1})[11]:

Formation of the H-bond system between considered polymers was confirmed fist of all by noticeable lowering (in 6 cm^{-1}) of the ν_{C-O-C} vibration band of PEO in the polymer mixture (Figure 6 a). Moreover, the contribution α of the band of 1723 cm^{-1} and also the effective length of *trans*-multimers of amide groups in the polymer mixture sharply increases (Table 6). Such effects can be attributed to appearance of the continuous sequence of H-bonds between polymer components.

Conclusion

The effect of self-assembly of partially destroyed aggregates of the short PVA chains on PAA long macromolecules as on matrices is established. At mixing of PVA with PAA in aqueous medium InterPC of the constant composition φ_{char} is formed. The value φ_{char} does not depend on the polymer component ratio and is determined mainly by molecular parameters of PAA. Disaggregating of PVA and PAA macromolecules in the process of complex formation has been revealed.

PEO with relatively short polymer chains (M ≤ 4·10^3) weakly interacts with PAA. Polymer-polymer interaction is intensified when the part of amide groups on PAA chain is transformed in the carboxylic groups. InterPCs formed have associative structure with friable packing of segments of both polymer components.

[1] V. A. Kabanov., *Vysokomolek. Soed.* **1994**, *A36*, 183.
[2] V. A. Izumrudov, A. B. Zezin, V. A. Kabanov, *Usp. Khim.* **1991**, *60*, 570.
[3] G.A. Mun, Z.S. Nurkeeva, V.V. Chutoryansky, R.A. Mangazbaeva, *Vysokomolek. Soed.* **2001**, *B43*, 552.
[4] D.V. Pergushov, V.A. Izumrudov, A.B. Zezin, V.A. Kabanov, *Vysokomolek. Soed.* **1995**, *A37*, 1739.
[5] Ukr. P. 17242 A; Promislova Vlastnist'. **1997**.
[6] N.M. Permyakova., T.B. Zheltonozhskaya, O.V. Demchenko, L.N. Momot, S.A. Filipchenko, N.E. Zagdanskaya, V.G.Syromyatnikov, *Polish J. Chem.* **2002**, *76*, 1347.
[7] V.S. Skazka, *Sedimentatsionno-diffuzionniy analys polymerov*, LGU, Leningrad **1985**, p. 251.
[8] A.D. Antipina, I.M. Papisov, V. A. Kabanov., *Vysokomolek. Soed.* **1970**, *B12*, 329.
[9] G. Nurberger, *Approximation by Spline Function, Springer,* Verlag **1989**, 243.

[10] A. Dunkan, *Prymeneniye Spectroskopii v Khimii, Izd-vo inostr.lyter*, Moskva **1959**, 159.
[11] T.V. Vitovetskaya, Thesis in Chemical Sciences, Kyiv, National University, Kyiv, **2003**.

Intramolecular Polycomplexes Formed by Dextran-Grafted Polyacrylamide: Effect of the Components Molecular Weight

Nataliya Kutsevol,[1] *Jean-Michel Guenet,*[2] *Nataliya Melnik,*[3] *Claudio Rossi*[3]

[1]Kyiv Taras Shevchenko National University, Department of chemistry, 60 str. Volodymyrska, KIEV 01033, Ukraine
E-mail: kutsevol@ukr.net
[2]Institute Charles Sadron, ULP-CNRS, 6 rue Boussingault, BP 40016 STRASBOURG, 67083, Cedex, France
[3]University of Siena, Department of Chemical and Biosystem Science, via Aldo Moro 2, SIENA 53100, Italy

Summary: Graft copolymers of Polyacrylamide-g-Dextran were synthesized using ceric-ion-initiated solution polymerisation technique. It is shown that the macromolecular parameters of these copolymers and their molecular structure strongly depend on the molecular weight of the Dextran part.

Keywords: dextran; graft copolymer; intramolecular bond; polyacrylamide

Introduction

Graft copolymerisation is one of the techniques employed for modifying the chemical properties of polymer. In the last decade in-depth study has been made on the synthesis and application of graft copolymers based on natural and synthetic polymers as these copolymers are very efficient flocculants [1]. A number of copolymers have been synthesized by grafting polyacrylamide (PAA) onto starch [2], carboxymethyl cellulose [2], guar gum [3], xanthan gum [4], and s.o.. It was concluded that by grafting flexible PAA chains onto the polysaccharide backbone, it is possible to develop efficient, shear-stable polymers for treatment of industrial effluents and for mineral processing. So far, systematic characterization of such copolymers is missing, although its knowledge is very important for understanding the mechanism involved in the capture of pollutants. Clearly, the copolymer microstructure will have a direct bearing upon the final properties.

This paper is devoted to the synthesis and the characterization of water-soluble graft copolymers based on Dextran and Polyacrylamide. These systems should be interesting due their propensity to form hydrogen bonds between the main and the grafted chains as well as between grafts leading to various and unusual inter and intramolecular structure [5].

Experimental

Dextran (Leuconostoc) with characteristics: M_w=500 000 g/mol (D500) was obtain from Fluka. Dextrans with characteristics: M_w=20 000 g/mol (D20) and M_w=70 000 g/mol (D70) were obtain from Serva (Sweden). Ammonium cerium (IV) nitrate (CAN) from Aldrich (USA) was used as initiator. Acrylamide (AA) was obtain from Reanal (Hungary). It was recrystallized from chloroform and dried under vacuum at room temperature for 24 h.

Synthesis of Dextran-graft –Polyacrylamide copolymers

0.02 mmol of Dextran was dissolved in 100 ml of distilled water. The contents of the flask were stirred and bubbled with argon gas for 20 min to remove the dissolved oxygen. Then, 0.12 mmol of CAN in 4 ml of 0.125 N HNO_3 was added to reaction mixture and allowed to react for 10 min, followed by the addition of calculated amount of AA. The polymerisation proceeded under argon atmosphere generally for 24 h. At the end of the reaction, the polymer was precipitated by adding an excess of acetone, dissolved in water again and then freeze-dried.

The elemental analysis of all samples was carried out for the four elements, i.e. carbon, hydrogen, oxygen and nitrogen.

The FT-IR spectra were obtained by Spectrophotometer Nicolet NIXUS-475 (USA) in the range 4000-400cm^{-1} using KBr pellets.

H^1 NMR spectra were recorded on an AVANCE spectrometer equipped with an xyz gradient unit, operating at 600.13 MHz. For the H^1 NMR measurements, the samples were run in D_2O. Viscosity measurements were performed for dilute solution in a bath thermostated at 25±0.05°C, using an Ostwald type viscometer. The data were analyzed according to:

$$\eta_{sp}/C = [\eta] + k_H[\eta]^2 C$$

where η_{sp} is the specific viscosity, $[\eta]$ the intrinsic viscosity and k_H the Huggins constant.

Results and Discussion

Two series of graft copolymers Polyacrylamide grafted to various molecular weight of Dextran (D20, D70, D500) named D20-g-PAA1, D70-g-PAA1, D500-g-PAA1 – *I series* and D20-g-PAA2, D70-g-PAA2, D500-g-PAA2 –*II series* were synthesised. The AA monomer concentration for copolymers of I and II series differed twice for getting various length of grafted chains.

For characterization of graft copolymers molecular parameters we calculated the percentage of PAA and Dextran in graft copolymers, ratio of PAA and Dextran monomers in copolymers β ($\beta=\alpha_1/\alpha_2$), where α_1 is the number of PAA monomers in macromolecules, α_2 the number of Dextran monomers in macromolecules of graft copolymer) from the data of elemental analysis. Take into account molecular weight of Dextran (M_D), the total molecular weight of all PAA grafts M_{PAA}, the molecular weight of the graft copolymers M_G ($M_G= M_{PAA}+M_D$) was calculated. It has been found also a significant percentage of water for all samples. The results are gathered in Table 1.

Table 1. Molecular characteristics of graft copolymers calculated from the elemental analysis.

Series	Sample	β	$M_D \times 10^{-3}$	$M_G \times 10^{-6}$	$M_{PAA} \times 10^{-6}$	Moisture, weight %
	D20-g-PAA1	31	20	0.41	0.39	9.22
I	D70-g-PAA1	20	70	0.69	0.62	8.23
	D500-g-PAA1	3	500	1.11	0.61	8.13
	D20-g-PAA2	45	20	0.28	0.26	8.21
II	D70-g-PAA2	11	70	0.41	0.34	9.14
	D500-g-PAA2	2	500	0.92	0.42	9.79

The total molecular weight of grafted PAA chains within I and II series is different. With increasing the molecular weight of Dextran in the graft copolymer also increases the total molecular weight of PAA component, however the ratio of total molecular weight of PAA grafts in the corresponding copolymers (based on the same molecular weight Dextran) of the I and II series is approximately equalled: for samples D20-g-PAA -1.5, for D70-g-PAA - 1.82 and for D500-g-PAA - 1.45.

In IR spectrum of D70-g-PAA (Fig.2), one can also observe the characteristic Amid I (stretching vibration of C=O groups) and Amid II (deformation vibration of NH_2 groups) bands of PAA at 1730-1580 cm^{-1} in addition to the characteristic absorption bands of Dextran glucosidic rings at 600-950 cm^{-1}, 1100-1200 cm^{-1} [6,7]. This is consistent with the occurrence of graft copolymerisation.

The chemical structure of graft copolymer was further examined by ^{1}H-MNR spectroscopy. A typical example of a ^{1}H-MNR spectra of the samples D20-PAA2 is drawn in Fig. 3. The characteristic resonance peaks corresponding to the protons in the methylene groups and other five methine groups of dextran at 3.40-3.92 ppm and 4.91 ppm [8] are apparently presented. The weak strip intensity of Dextran components in graft copolymers (Fig.3) is explained by the low content of Dextran in this samples – 6.35%, calculated from the elemental analysis data. In addition, the signals of protons in PAA graft also appear in the spectrum at 2.2-2.3 ppm (methine groups), 1.6-1.7 ppm (methylene groups), which a close to experimental results, has been obtain in [9] for individual PAA. These results also testifies the existence of grafting.

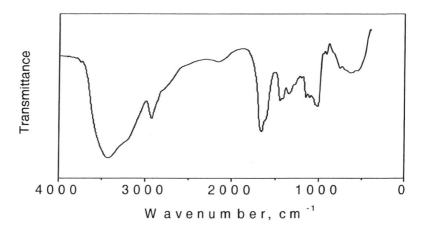

Figure 1. FTIR spectrum of D500-g-PAA.

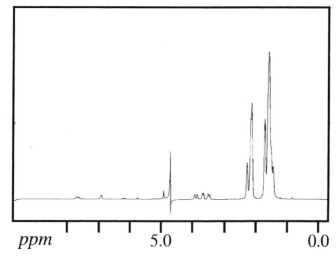

Figure 2. ^{1}H-NMR spectrum of D20-g-PAA2 copolymer.

Figure 3 shows the plots of the reduced viscosity (η_{sp}/C) versus concentration for D500-g-PAA2, D70-g-PAA2 and D20-g-PAA2. Through extrapolation to zero concentration the intrinsic viscosity [η] of the graft copolymers was obtained (Table 2).

Table 2. Molecular parameters of graft copolymers determined by viscometry and elemental analysis methods.

Series	Sample	$M_G \times 10^{-6}$	[η], m^3kg^{-1}
I	D20-g-PAA1	0.41	0.40
	D70-g-PAA1	0.69	0.65
	D500-g-PAA1	1, 11	0.51
II	D20-g-PAA2	0.28	0.23
	D70-g-PAA2	0.41	0.20
	D500-g-PAA2	0.92	0.31

We did not observe a drastic change in intrinsic viscosity for copolymers within I and II series in spite of the differing molecular weight of these copolymers (Table 2). Taking into account that intrinsic viscosity characterizes the hydrodynamic volume of the coil in solution, we can therefore conclude that the structure of graft copolymer D-g-PAA depends strongly on the molecular weight of Dextran component. This may arise from the possibility of Dextran and PAA components to form intramolecular H-bonds. Parameter β in Table 1 characterizes the ratio of Dextran and PAA monomers in the copolymers. The smallest value of this parameter is for copolymers based on D500. This suggests that in this case the bonding between the main chain and the grafted chains is more efficient.

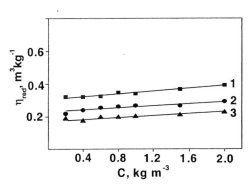

Figure 3. Dependence of reduced viscosity vs concentration for D500-g-PAA2 (1), D70-g-PAA2 (2) and D20-g-PAA2 (3).

The temperature dependence of the reduced viscosity for D70-g-PAA1(a) and D500-g-PAA2 (b) is shown in Fig.4. Significantly differing types of behaviour are seen. This also suggests that the intramolecular bonding varies with the number of grafts.

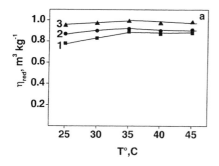 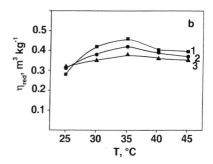

Figure 4. Temperature dependence of reduced viscosity for D70-g-PAA1(a) and D500-g-PAA2 (b). C=0.01 (1), 0.02 (2), C=0.6 kg m^{-3} (3).

Conclusion

It can be concluded that the macromolecular structure of D-g-PAA copolymers strongly depends on the molecular weight of Dextran component. The structure of such compounds can be easily monitored through the ratio of the main chain length to the grafts length.

[1] T.T. Tripathy, R.P Singh. *Eur.Pol. J.* , **2000**, *36*, 1471.

[2] S.R. Deshmukh, K. Sudhakar, R.P. Singh. *J. Appl. Polym. Sci.*, **1991**, *43*, 1091.

[3] S.R. Deshmukh, R.P. Singh. J. *Appl. Polym. Sci.*, **1987**, *33*, 1963.

[4] S.R. Deshmukh, R.P. Singh. *J. Appl. Polym. Sci.*, **1986**, *32*, 6163.

[5] V. Syromyatnikov, T. Zheltonozhskaya, O. Demchenko, J.-M. Guenet, I. Rakovich, Strelchuk, N. Permyakova. *Macromol. Symp.*, **2001**, *166*, 237.

[6] V.D. Athawale, V. Lele. *Carbohydrate Pol.*, **1998**, *35*, 21.

[7] P. Ghosh, D. Dey. *Eur. Pol. J.*, **1996**, *32*, 165.

[8].L.-Q. Wang, K. Tu, Y. Li, J. Zhang, L. Jiang, Z.Zhang. *React. & Func. Pol.*, **2002**, *53*, 19.

[9] O. Demchenko, T. Zheltonozhskaya, A. Turov, M. Tsapko, V. Syromyatnikov. Abstracts of 5th Conf. "*Electronic processes in organic materials*", May 24-29, **2004**, Kyiv, Ukraine, p.176.

Dynamics and Crystallization in Polydimethylsiloxane Nanocomposites

Alexandre Beigbeder,[1] *Stéphane Bruzaud,*[1] *Jiri Spěváček,*[2] *Jiri Brus,*[2] *Yves Grohens**[1]

[1]Laboratory of Polymers, Properties at Interfaces & Composites, South Brittany University, Rue de Saint-Maudé, 56 321 Lorient, France
E-mail: yves.grohens@univ-ubs.fr
[2]Institute of Macromolecular Chemistry, Academy of Sciences of the Czech Republic, 16206 Prague 6, Czech Republic

Summary: Several routes were used to achieve silicon nanocomposites. The first and second one are the melt intercalation of polydimethylsiloxane (PDMS), which is a mechanical blending of the polymer in the molten state with the untreated inorganic filler or intercalated nanoparticles. The last one is an in situ polymerization, which previously requires the intercalation of hexamethylcyclotrisiloxane (D_3) followed by a subsequent polymerization step. We used synthetic mineral oxide $HTiNbO_5$ as nanofiller. These systems were investigated by differential scanning calorimetry (DSC) and solid state NMR in order to better understand the relation between the nanocomposites dynamics, and crystallisation. The efficiency of grafting reactions was studied by ^{29}Si CP/MAS NMR. The nature of the interfacial interactions seems to play the major role. Indeed, the nanocomposites 1 and 2 for which only physical interactions are expected do not exhibit any T_g deviation whereas the nanocomposite 3, for which chemical grafting is achieved, increases strongly the T_g. Crystallization is more sensitive to density and strength of interfacial interactions which are maximum for the pristine filler.

Keywords: crystallisation; molecular dynamics; polydimethylsiloxane nanocomposites

Introduction

During this last decade, much attention has been paid to polymer nanocomposites especially polymer-layered silicate nanocomposites, which represent a rational alternative to conventional filled polymers. The most often used nanofillers are modified layered clays such as montmorillonite, bentonite and things like that. Because of their nanometric size and their large active surface, it can be expected that polymeric nanocomposites exhibit improved mechanical, thermal, dimensional and barrier properties compared to pure

© 2005 WILEY-VCH Verlag GmbH & KGaA, Weinheim   DOI: 10.1002/masy.200550429

polymers. [1-4] In our approach, we used $HTiNbO_5$ synthetic mineral oxide which was chosen for its perfect lamellar structure and for its well-defined chemical structure, contrary to layered silicate clays. Furthermore, several studies have demonstrated the great flexibility of $HTiNbO_5$ structure and the possibility of intercalation of voluminous organic molecules. [5-7] In this work, we studied different polysiloxane-g-$TiNbO_5$ nanocomposites by solid state differential scanning calorimetry (DSC) and solid state nuclear magnetic resonance (NMR). For the synthesis of these nanocomposites, three routes were investigated, which are mechanical blend of PDMS matrix and $HTNbO_5$ or treated $HTNbO_5$ (intercalated by tetramethylammonium hydroxide) and in-situ anionic ring-opening polymerization of cyclosiloxanes.

Experimental part

The elaboration of nanocomposites is explained is previous papers. [8-9] Nano 1 and 2 are prepared by blending the mineral and the polymer. For Nano 3, the initial product is composed by 40% of mineral and 60% of polymer and the obtained hybrid material is mixed with pure PDMS ($Mn = 4000$ g/mol) to achieve the desired filler amount (Table 1).

Table 1. Composition and characterization (WAXS and rheology) of the various nanocomposites.

Sample	Mineral	Monomer	Process	WAXS and rheology
Nano 1	$HTiNbO_5$	PDMS	Mechanical blend	Layer organised structure not intercalated by PDMS chains
Nano 2	$(Me_4N)_xH_{1-x}TiNbO_5$ $x \sim 0.4$	PDMS	Mechanical blend	Mineral intercalated by the surfactant but not by PDMS chains
Nano 3	$(Me_4N)_xH_{1-x}TiNbO_5$ $x \sim 0.4$	Hexamethylcyclotrisiloxane D_3	Ring opening polymerization	Mineral partially exfoliated

The crystallization and glass temperature characterization was carried out in a differential scanning calorimeter (DSC 822 METTLER TOLEDO) with a disk-type measuring system and all the heating heating runs were done at 10°C/min. The glass transition temperature was determined as the temperature corresponding to the inflection point of the glass transition step. The temperature of the crystallisation peak corresponds to the maximum of the heat flux after normalisation correction.

^{29}Si magic angle spinning (MAS) NMR spectra were measured on a Bruker Avance 500 solid state NMR spectrometer at 99.37 MHz with spinning frequencies 3-5 kHz. ^{29}Si NMR spectra were measured both with or without cross polarization (CP). In ^{29}Si CP/MAS NMR spectra the contact time 3 ms and relaxation delay 10 s were used. ^{29}Si spin-lattice relaxation times T_1 were measured by the method analogous to that of Torchia[10], but without CP. Relaxation delay in T_1 measurements was set to be at least 5 times longer than the respective T_1 (80-200 s depending on the sample and temperature). The measurements were done on nanocomposites with 2% filler content.

Results and discussion

Table 2 shows the crystallisation temperature and exothermic peak of Nano 1, 2 and 3 as a function of the mass fraction of nanofiller present in PDMS matrix.

Table 2. Crystallisation characteristics of nanocomposites 1, 2, and 3.

Nanocomposite	Mass fraction of filler (%)	Crystallisation temperature (°C)	Enthalpy (J/g)
Nano 1	0	-88	21.8
	2	-100	4
	5	-100	0.3
	7	No peak	----
	10	No peak	----
Nano 2	0	-88	21.8
	2	-94	8.3
	5	-99	1.3
	7	-97	----
	10	No peak	----
Nano 3	0	-88	21.8
	2	-92	9.4
	5	-93	6.3
	7	-94	4.7
	10	No peak	----

Crystallization is one of the most effective processes used to control the extent of intercalation of polymer chains into mineral galleries, and hence to control the mechanical and various other properties of the nanocomposites[11]. The crystallization temperature T_c changes in average from –88°C to –100°C for Nano 1, to –96 for Nano 2 whereas it only reaches -93°C for Nano 3 system. The nanoparticles act as nucleating agents by promoting the crystallization at lower temperatures. Moreover, the enthalpy of crystallization strongly decreases whatever the system and this decrease is ranked as follows: 1 > 2 > 3. This is due to a lowering of the size of the spherulite dimensions as claimed by other authors for different systems[11]. The adsorption of PDMS chains on the filler can also partially hinder the ability of some bonded chains to self organize in a crystallize lattice.

The main difference between our systems is the chemistry of the mineral surface: (OH groups for Nano 1, intercalated tetramethylammonium ions for Nano 2, and PDMS chains (resulting of D_3 polymerisation) for Nano 3. The nature of the interfacial interactions seems to play a significant role. Crystallization seems to be sensitive to the density of interfacial interactions, which are direct mineral/PDMS contacts for Nano 1, which yield the higher lowering of T_c and crystallization enthalpy. Alkylammonium mediated interactions for Nano 2 and grafted PDMS chains mediated interactions for Nano 3 yield lower depression of T_c and crystallization enthalpy.

Table 3 shows the glass transition temperature of the various systems obtained from DSC thermograms. Only the Nano 3 system exhibits strong deviation from the pure bulk T_g which increases from –125°C to –96°C for the highest mineral amount. Therefore, chemically grafted chains exhibit a very different cooperative behavior than physisorbed chains. The strong enlargement of the transition temperature range (14°) is often ascribed to confinement effects in nanocomposites [12].

The DSC thermograms of the other systems do not show any appreciable change of T_g or an hindered T_g with the increasing $HTiNbO_5$ concentration. This would indicate that the presence of the mineral doesn't produce important changes in the mobility of the polymer chains for 2% of mineral. For higher mineral content the unrecorded T_g may be due to an hindered cooperativity in the motions of the chains according to their very different environment and conformation. This is consistent with the assumption of a ''frozen layer'' of polymer in contact with the surface [14]. The very low ΔC_p (0.006) for Nano 1 at 2% accounts for this assumption.

Table 3. Effect of nanofiller mass fraction on T_g, ΔC_p and ΔT.

Nanocomposites	Mass fraction (%)	T_g (°C)	ΔC_p (J.g^{-1}.K^{-1})	ΔT (°C)[a]
PDMS		-125.4	0.447	5,7
Nano 1	2	-125.6	0.006	4,7
	5-7-10	no T_g	----	----
Nano 2	2	-125.9	0.178	5,3
	5-7-10	no T_g	----	----
Nano 3	2	-124.5	0.213	7,4
	5	-125.2	0.170	5,7
	7	-125.0	0.147	5,7
	10	No T_g	----	----
	40	-96.0	0.191	14,2

[a]ΔT = *Temperature range of the glass transition phenomena*

The ^{29}Si NMR relaxation times T_1 were measured for Nano 1,2,3 as well as for the polydimethylsiloxane (PDMS) matrix. Moreover, measurements of ^{29}Si NMR spectra with cross-polarization (CP) were carried out for all samples and the results are shown in Table 4 and Figure 1. The values of T_1 increase from 9.5 s to a maximum of 35.6 s and that for few percent of nanofiller (2%). Surprisingly, the relaxation times of the all the Nano samples are higher than that for the neat PDMS matrix. From T_1 measurements at elevated temperatures it follows that the mobility of the chain segments at the molecular level is higher when nanofiller is introduced. This is not in contradiction with the previously observed increase of T_g or ''frozen layer'' which is relevant of larger scale cooperative motions. However, the exact reason for this higher mobility is not well understood. The organization of the chains at the mineral surface is probably very different in comparison with that in the bulk. The lower entanglement density in interfacial regions, which was claimed by several authors [13], is not consistent with the low molecular weight of our PDMS. The only possible explanation is an increase of the free volume due to a lower packing density at the filler surface in comparison with the bulk. No difference is observed for T_1 according to the nature of the PDMS/surface interaction, namely, physi- or chemisorption.

230

Table 4. ^{29}Si NMR relaxation times T_1 of various systems at 27 °C.

System	Relaxation time T_1 (s)
PDMS	9.5 (19.4 at 37 °C)
Nano 1	27.9
Nano 2	35.6
Nano 3	33.7

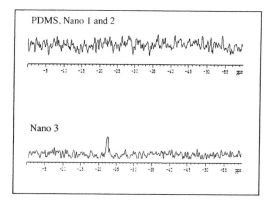

Figure 1. ^{29}Si NMR spectra with cross-polarization

No signal in ^{29}Si NMR spectra measured with CP was detected for the neat PDMS and Nano 1 and 2 (Figure 1), due to high segmental mobility. For Nano 3, a weak signal was obtained at −22 ppm which reveals the existence of near-static dipolar interactions for a small portion of PDMS units in this sample. It can be assumed that this signal is an evidence for the existence of the chemical grafting of PDMS chains; it evidently corresponds to PDMS units in the vicinity of the graft points. Indeed, the ring opening polymerization of D_3 leads to the formation of covalent bonds with the surface anions. However the weak intensity of this signal (in comparison with intensity in ^{29}Si NMR spectra measured without CP) accounts for a rather low grafting density.

Conclusion

PDMS nanocomposites were elaborated by several routes yielding different structure, surface organization and thermal properties. No correlation was found between the state of dispersion of the nanofillers in the PDMS matrix determined by rheology and both the T_g and crystallization of the nanocomposite. The nature of the interactions between polymers and mineral seems to play an essential role. The weak signal in ^{29}Si CP/MAS NMR spectrum was ascribed to PDMS chain grafting onto the HTiNbO$_5$ surface. Finally, T_g is modified for PDMS grafted chains whereas the higher changes in the crystallization rate are achieved for strongly physisorbed chains.

Acknowledgment

Support in the frame of the EU IHP program (HPMT-CT-2001-00396, a Marie Curie fellowship for A. B.) is gratefully acknowledged.

[1] S.D Burnside, E. P. Giannelis, *Chem. Mater.* **1995**, *7*, 1597.
[2] E. P. Giannelis, R. Krishnamoorti, E. Manias, *Adv. Polym. Sci.* **1999**,*138*, 107.
[3] M. Alexandre, P. Dubois, *Mater. Sci. Eng.* **2000**, *R28*, 1.
[4] K. E. Strawhecker, E. Manias, *Chem. Mater.* **2000**, *12*, 2943.
[5] H. Rebbah, G. Desgardin, B. Raveau, *Mater. Res. Bull.* **1979**, *14*, 1125.
[6] H. Rebbah, M. M. Borel, B. Raveau, *Mater. Res. Bull.* **1980**, *15*, 317.
[7] S. Kikkawa, M. Koizumi, *Mater. Res. Bull.* **1980**, *15*, 533.
[8] S. Bruzaud, G. Levesque, *Chem. Mater.* **2002**, *14*, 2421.
[9] A. Beigbeder, S. Bruzaud, T. Aubry, P. Médéric, Y. Grohens, *Polymer* (to be published).
[10] D.A.Torchia, *J. Magn. Reson.* **1978**, 30, 613.
[11] P. Maiti, P. H. Nam, M. Okamoto, T. Kotaka, N. Hasegawa, A. Usiki , *Macromolecules* **2002**, *35*, 2042.
[12] S.Vyazovkin, I.Dranca, *J.Phys.Chem.B* **2004**, 108, 11981
[13] Forrest J.A., Danolki Veress K., *Adv. Colloid Interface Sci.* **2001**, 94, 167

Elaboration and Characterization of Starch/ Poly(caprolactone) Blends

Isabelle Pillin, Thomas Divers, Jean-François Feller, Yves Grohens*

Laboratoire Polymères, Propriétés aux Interfaces et Composites, Université de Bretagne-Sud, Rue de Saint-Maudé, BP 92 116, 56321 Lorient, France
Fax: (33) 2 97 87 45 88; E-mail: yves.grohens @univ-ubs.fr

Summary: A starch-based biodegradable material was prepared in two steps. Firstly, starch was chemically modified by using formic acid at 20°C to obtained degrees of substitution of about 1.2. The level of destructuration was also assessed using dynamic rheological measurements. Native starch or starch ester were then mixed with poly(caprolactone) and different polyester oligomers were added as compatibilisers and plasticizing agents. PCL oligomers were found to be the most efficient ones. A significant improvement of the elongation at break of starch formate/PCL/oligo PCL blends was achieved.

Keywords: biodegradable blend; biodegradability; destructuration; O-formylation; PCL; rheology; starch

Introduction

The natural susceptibility of starch to biodegradation has sparked a considerable amount of research to provide starch-containing materials able to undergo rapid biologically induced destruction. Blends of native starch and poly(olefin)s have been largely studied and lead to heterogeneous materials with poor mechanical properties. This low cohesion results from the weak adhesion between starch (hydrophilic) and poly(olefin)s (lipophilic) [1,2]. Moreover, biodegradability of such blends is not complete due the presence poly(olefins). That is why other blends of starch with synthetic biodegradable polymers have been tested. Amongst the available polymers, poly(caprolactone) is one of the most studied. However, the resulting material has also weak properties when using native starch, what can also be improved by the use of compatibilisers or by modification of starch [3-7]. Nevertheless, to achieve this result it is first necessary to, at least partially, remove starch crystalline organization through swelling and gelatinization in water, which can be controlled adjusting the acidic character and the temperature of the medium [8]. Since the 1940's, O-formylation of starch has been widely studied [9-13]. Gottlieb *et al.* [12] have shown that the reaction of formic acid on starch induced the rapid formation of a monoformic ester (DS = 1). According to Wolff *et al.*[10], starch formylation is a reversible reaction where the

extent of substitution depends on the ratio of formic acid to starch. Recently, Aburto *et al.*
[9] have described a route of synthesis of starch fatty acid esters without an organic solvent.
At first, native starch is treated with formic acid and then, octanoyl chloride is added to
form octanoated starch.

In this paper, we present the destructuration of starch in formic acid and the preparation of
starch (formiate) /PCL/oligo polyesters blends and the study of their mechanical and
rheological properties.

Experimental

Wheat starch (I59-113H10) was provided by ROQUETTE (France), that is composed of
25% amylose, 75% amylopectin and has a molecular weight of about 50.10^6 g.mol^{-1}.
Water content was measured by TGA and was found to be 13%. Starch samples were
dried at 105 °C for 4h before use to a water content close to 1 %. Formic acid solution 99
% (FA) was used as received from Sigma Aldrich. PCL CAPA 6800 was provided by
SOLVAY. Its molecular weight is about 80,000 g.mol^{-1}, density of 1.11g.cm^{-3}. Polyester
prepolymers were provided by DUREZ (table 1).

Table 1. Main characteristics of DUREZ prepolymers.

Oligomer	Nature	Functionnalization	Mw (g.mol^{-1})
P1 : 105-42	1,6- hexane-diol adipate and phthalate	Hydroxyl	2700
P2 : 101-55	Glycol adipate and phthalate	Hydroxyl	2000
P3 : 105-15	1,6-hexane-diol adipate and phthalate	Hydroxyl	7400
P4 : 1063-35	Polycaprolactone	Hydroxyl	2000

50 grams (0.31 mol) of dry native starch were introduced in a three-necked flask
containing 250 mL of 99 % formic acid (FA, 6.62 mol). The mixture was stirred at 20°C
for 6 hours. Then, the solution was gently poured into methanol (1 L) and filtered off and
washed three times with methanol (3 × 300 mL) to remove FA in excess. The samples
were then dried in an oven at 50 °C for 24h under vacuum. Degrees of substitution were
determined as described elsewhere [14] and measured of about 1.2.

The blends were extruded with a twin screw extruder (Brabender, DSK 42/6) controlled by
a Lab-Station with a screw rotation speed of 30 rpm and temperatures of respectively 85,

90 and 95°C from hopper to slot die (4´50 mm^2).

Rheological properties of native starch and starch formate were determined using a THERMOHAAKE Rheostress 1 rheometer with the parallel plates geometry. The samples (native starch or starch formate, 10% (w/w)), were analyzed in dynamic mode with controlled strain. The frequencies range scanned were 0.05 to 50 Hz. All the experiments conducted were isothermal and the temperatures varied from 25°C to 80 °C.

Tensile tests were realized on a MTS Synergie RT1000, using standard ISO527.

Results and discussion

The reaction of starch with formic acid (FA) (Figure 1) was carried out in a thermally isolated calorimeter under mechanical stirring. The stoechiometry between formic acid and starch hydroxyl groups at the beginning of the reaction was 7.05/1.

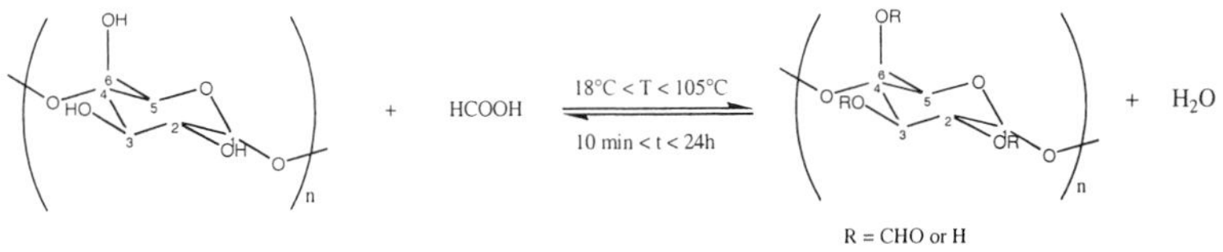

R = CHO or H

Figure 1. O-formylation reaction on starch.

In order to inhibit depolymerization and color formation, O-formylation was performed on wheat starch at 20 °C for 6 hours and DS measured is about 1.2. Due to their important sensitivity to molecular interactions and structures, rheological measurements were used to characterize starch destructuration during the esterification reaction [15,16]. In excess water, it has been shown that water enters amorphous regions first. Subsequently, it is where the granule swelling occurs. As the temperature increases, crystallites are destroyed, generating both a crystallinity and birefringence loss and leading to the creation of a three dimensional network [17].

In pure formic acid, starch gelatinization proceeds at room temperature (Figure 2). In fact, gelatinization in formic acid occurs at lower temperatures than in water (78°C). Surprisingly, when Tr is increased up to 50 °C, the independent behavior of macromolecules observed at 40 °C disappears and instead, a cooperative behavior is evidenced, indicating a new gel structure. The new gel is a stiffer network than the one at 25-30 °C. This phenomenon may be explained by a competition between chain scissions due to hydrolysis leading to a higher mobility and an aggregation process of chains resulting from an unknown mechanism but likely due to O-formylation.

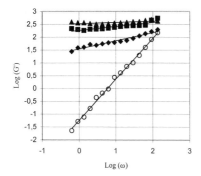

Figure 2. Evolution of the storage modulus (G') with frequency as a function of temperature for native wheat starch suspension in 99% (10% mixture of wheat starch in FA) (gap: 0.3 mm, $\gamma_0 = 8\%$). ◆ represents G' evolution at T = 25°C, ■ at 30°C, ○ at 40°C and ▲ at 50°C.

Different hydroxyl functionnalized polyester oligomers have been used as compatibilisers of starch/PCL blends. In table 2, we can first notice that starch formate provides blend with PCL with higher modulus than for native starch (Formulations F1 and F2). Therefore, starch acts as a reinforcing agent on PCL which itself exhibit a very low modulus. However, only even for starch formate weak miscibility is expected from the strong reduction of the strain and stress at the rupture. Large starch aggregates are present in the blend and this lack of homogeneous dispersion yields the fragile behaviour observed for the blends. Plasticizing agents have already been claimed by other workers [3-7] and are the aim of the telechelic polyester oligomers that have been studied here.

In comparison with pure PCL/starch blends, the stress at break of PCL/oligomers/starch blends is reduced whereas elongation at break can be enhanced (formulation F1 compared to formulations F3 to F6). The less miscible oligomer (prepolymer P2) yields a material which is very brittle (strain at break 0.4%) without any cohesion. The increase of the molecular weight of the oligomer increases the stress and strain at break in that formulation (F5) which is consistent with the expected behaviour.

PCL oligomer (F6) yields an elongation at break of 191% which is much higher than all the others. This is due to the good miscibility of the PCL oligomer with the PCL but rather strong interaction are also suspected with starch. Moreover, comparing formulations F6 and F7, we can observe that use of formate starch further increase the elongation at break (from 191 to 247%) in comparison to native starch.

Table 2. Melt index flow and mechanical tensile properties of blends containing starch (or formate starch), oligomer and PCL CAPA 6800 40/30/30 (w/w).

Formulations	Starch	Formate starch	P1	P2	P3	P4	PCL	MVI (cm³/10min)	Module (MPa)	Stress at break (MPa)	Strain at break (%)
Pure PCL							100	8.0	262.0	31.5	693.0
F1	40						60	4.7	655.0	8.9	6.2
F2		40					60	14.0	1095.0	10.1	1.4
F3	40		30				30	18.2	1002.0	2.8	3.6
F4	40			30			30	541.0	828.0	1.7	0.4
F5	40				30		30	22.8	832.0	7.8	11.8
F6	40					30	30	27.1	280.0	5.1	191.0
F7		40				30	30	38.0	266.0	4.0	247.0

This phenomenon, observed only with this oligomer, let us suppose that strong affinity could exist between starch formate and oligomer PCL. In order to verify this hypothesis, different formulations were tested with native or modified starch (DS=1.2). Rheological investigations carried out at 90°C showed that starch formate exhibit a different rheological behaviour in the blends than native starch. The frequency independence of G' and G'' for starch formate blends is the sign of a gel formation which, in turn, is ascribed to a strong destructuration of starch due to strong interactions with PCL oligomer (Figure 3).

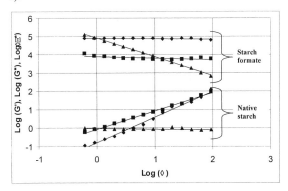

Figure 3. Evolution of the storage and loss moduli (respectively G' and G'') and the complex viscosity as a function of the frequency for native and modified starch mixed with PCL oligomer. (gap: 0.3 mm, g0 = 8%, T = 90°C). ♦ represents G' evolution, ■ G'' evolution and ▲ η* evolution).

Indeed, a gel behavior is deduced from the frequency independent behavior of G' and G''
in the case of starch formate as shown in figure 3. This is not observed for wheat
starch/PCL blend which behaves as a dispersion of starch granules in PCL:
Destructuration is not sufficient without a high shear at this temperature. This is consistent
with the observation made by other groups [5,6]. This rheological observation yields the
conclusion of a better affinity of modified starch for PCL than wheat starch. These strong
interactions may mostly be due to intermolecular H-bonds oligo PCL/starch formate which
substitute amylopectine/amylose interactions.

Conclusion

Storage modulus (G') values measured by dynamic rheometry were found to be sensitive
to structural changes of starch as a function of time and temperature of reaction. It was
shown that the slope of G' versus frequency (w) was a pertinent parameter to follow both
gelatinization and gel destruction and to determine the temperature domain in which the
starch has a gel structure in formic acid. Concerning the blends of starch with PCL, it has
been shown that O-formylation reaction allows to increase interaction between starch and
PCL and that PCL oligomers can act as a platicizing agent to significantly increase the
elongation at break of the biodegradable blends.

Acknowledgements

The authors would like to gratefully thank M. Lénaïck Lemée, Ms. François Péresse, M.
Bruno Perly and the ROQUETTE society for the wheat starch. This project was supported
by the Brittany Region and the French Ministry of Research & New Technologies.

[1] S. T. Lim, J. L. Jane Rajagopalan, P. A. Seib, *Biotechnol. Prog.,* **1992**, *8*, 51.
[2] G. J. L. Griffin, *U.S. Patent 4016117,* **1977.**
[3] C.S. Wu:. *Polym. Deg. Stab.,* **2003**, 80, 127.
[4] R.P. Singh, J.K. Pandey, D. Rutot, Ph. Degée, Ph. Dubois, *Carbohydr. Res.,* **2003**, *338*, 1759.
[5] L. Averous, L. Moro, P. Dole, C. Fringant, *Polymer,* **2000**, *41*, 4157.
[6] M. Avella, M.E. Erico, P. Laurienzo, E. Martuscelli, M. Raimo, R. Rimedio, *Polymer,* **2000**, *41*, 3875.
[7] M.F. Koenig, S.J. Huang, *Polymer,* **1995**, *36*, 1877.
[8] O. Sevenou, S.E. Hill, I.A. Farhat, J.R. Mitchell, *Bio. Macromol.,* **2002**, *31*, 79.
[9] J. Aburto, I. Alric, E. Borredon, *Starch/Stärke,* **1999**, *51*, 132.
[10] I. A. Wolff, D. W. Olds, G. E. Hilbert, *J. Am. Chem. Soc.,* **1957**, *79*, 3860.
[11] H. Tarkow, A. J. Stamm, *J. Phys. Chem.,* **1951**, 56, 266.
[12] D. Gottlieb, C. G. Caldwell, R. M. Hixon, *J. Am. Chem. Soc.,* **1940**, *62*, 3342.
[13] I. A. Wolff, D. W. Olds, G. E. Hilbert, *J. Am. Chem. Soc.,* **1951**, *73*, 346.
[14] O. B. Wurzburg, "Acetylation", in *Methods in Carbohydrate Chemistry,* Ed Whistler, New York,
Academic Press, **1964**, 287.
[15] I. Rosalina, M. Bhattacharya, *Polymer,* **2002**, *48*, 191.
[16] B. Jauregui, M. E. Munoz, A. Santamarialnt, *J. Biol. Macromol.,* **1995**, *17*, 49.
[17] P. J. Jenkins, A. M. Donald, *Carbohydr Res.,* **1998**, *308*, 133.

Influence of Phase Morphology on Molecular Mobility of

Poly(propylene)-(ethylene-vinyl acetate) Copolymer Blends

Sylvie Pimbert,[*1] Isabelle Stevenson,[2] Gerard Seytre,[2] Gisele Boiteux,[2] Philippe Cassagnau[2]*

[1] Laboratoire Polymères, Propriétés aux Interfaces et Composites, EA 2592, Université Bretagne Sud, 56325 Lorient, France
E-mail: sylvie.pimbert@univ-ubs.fr
[2] Laboratoire Matériaux Polymères et Biomatériaux, UMR 5627, Université Claude Bernard, Bat ISTIL, 69622 Villeurbanne, France

Summary: The dynamic mechanical and dielectric behaviours of Polypropylene (PP) and (Ethylene-Vinyl Acetate) Copolymer (EVA) blends are reported as a function of the morphology. For EVA contents lower than 20%, blends show the two-phase morphology characteristic of immiscible blends, with spherical EVA droplets finely dispersed in the PP matrix. After stretching in the molten state, the morphology of EVA fibers is observed. Mechanical Relaxation Spectroscopy display three relaxation processes: the EVA and PP α-relaxations associated to the glass transitions and a β-transition corresponding to a PP crystalline phase relaxation. The PP α-relaxation shifts to higher temperatures when EVA presents a fiber morphology, corresponding to a decrease of PP chain mobility since it is hindered by the reinforcement effect of EVA fibers. Quite different results are obtained by DRS analysis. In blends containing EVA fibers, only one main relaxation associated to the EVA α-transition is observed whereas one additional relaxation can be noticed in the blends containing EVA droplets. This new relaxation might be assigned to interfacial polarization effects, phenomena that are sometimes observed in heterogeneous polymer blends when a low content of one polar component is embedded in a non conductive matrix. In this case, the occurrence of a characteristic interfacial polarization relaxation appears to be correlated to the accessible experimental frequency.

Keywords: dielectric relaxation spectroscopy; (ethylene-vinyl acetate) copolymer; mechanical relaxation spectroscopy; polymer blend morphology; polypropylene

Introduction

Nowadays, polymer blends are considered as one of the most important developments in polymer engineering because of the possibility of obtaining new materials with specific properties. An example is the system polypropylene/ poly(ethylene vinyl acetate) (PP/EVA) which combines two important commercial polymers : a semi-crystalline thermoplastic and an elastomer. This system has been the subject of several studies considering more particularly miscibility and compatibilization aspects [1-4]. Nevertheless, few papers are devoted to the correlation between the relaxation phenomena and the blend morphology in such heterogeneous systems [5-6]. More specifically, the interfacial polarization or Maxwell-Wagner-Sillars effect is scarcely described [7-8], particularly in polymer blends [9-10]. This paper presents a study of PP/EVA blends with low EVA contents (non miscible domain), considering the influence of blend morphology on molecular mobility. Dynamic and dielectric mechanical relaxation spectroscopy is used to investigate phase relaxations in relation to the conductive EVA phase behaviour.

Experimental

The materials investigated in this study are blends of polypropylene (iPP), as the main component, with an ethylene-vinyl acetate copolymer (EVA) as the dispersed phase. The PP supplied by ExxonMobil Chemicals is an Escorene 4352 F2 having a melting temperature of 163 °C, a specific gravity of 0.908 and a melt-index of 3 g/10min (ISO 1133). The EVA copolymer (EVATANE 28-03 from ATOFINA) has a melt-index of 3 g/10min (ISO 1133) and contains 28% vinyl acetate by weight. Blends were prepared with a twin screw extruder (Leistritz LSM 34 mm model, L/D=33.5) at 200 °C and 200 rpm with a 4 kg/h flow rate. EVA contents were respectively 2.5, 5, 10, 20 and 30 weight percent. For each composition, blends were extruded through a 50 x 2 mm slot die either without any drawing (pressed samples) or drawn uniaxially (drawn samples). DSC measurements were performed using a 2920 TA Instruments apparatus in the temperature range of -50 to 210 °C with a heating rate of 10 K.min^{-1} under nitrogen atmosphere. The temperature calibration was achieved using indium. Blend morphologies were examined with a HITACHI S800 Scanning Electronic Microscope (SEM) with a field emission gun, working at an acceleration voltage of 15 kV, on Au-Pd coated cryogenically fractured samples, after extraction of EVA with toluene.

Mechanical Relaxation Spectroscopy experiments were carried out using a DMA 2980 analyzer from TA Instruments, in the tensile film mode. The experimental conditions were a heating rate of 3 K.min^{-1}, a frequency of 1 Hz in the temperature range from -50 to 160 °C, using 500 μm thick films. For the dielectric measurements, circular aluminium electrodes (30 and 10 mm diameter) were evaporated on the top and the bottom of the sample. The complex dielectric function $\varepsilon^* = f(T, \nu)$ was measured as a function of temperature T (from -50 to 150 °C) and frequency (between 10^{-1} and 10^5 Hz) by a lock-in amplifier (Stanford Research 810) interfaced to the sample by a broadband dielectric interface (BDC, Novocontrol). The temperature T dependence of the α and β relaxations can be described by an Arrhenius equation: $f = f_1 \exp[-\frac{E_a}{kT}]$ which has allowed us to estimate the activation energy E_a (k is Boltzmann constant and f_1 is the pre-exponential factor).

Results and Discussion

Morphologies of PP-EVA blends were examined using SEM. Typical images are presented in Fig.1 and Fig.2 for samples without drawing and with uniaxial drawing respectively.

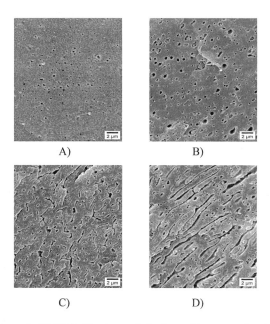

A)

B)

C)

D)

Figure 1. PP/EVA blend morphologies, without any drawing A: 95/5, B: 90/10, C: 80/20, D: 70/30.

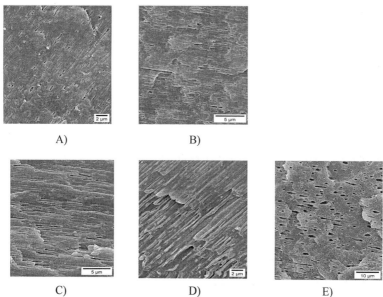

C)   D)   E)

Figure 2. PP/EVA blend morphologies, with uniaxial drawing A: 95/5, B: 90/10, C: 80/20, D: 70/30, E: 70/30 perpendicular to extrusion direction.

Micrographs obtained from samples without drawing show a two phase morphology of EVA droplets finely dispersed in the PP matrix for EVA content lower than 20%. This morphology is characteristic of immiscible blends. For higher contents (20-30%), co-continuity is reached with two interpenetrated phases. For samples uniaxially drawn at the die exit, a morphology of EVA fibers is observed. Their length increases with EVA content, up to 20% EVA. The PP-EVA 70/30 blend shows a fairly irregular and mixed morphology with larger EVA fibers, corresponding to the beginning of co-continuity.

In the composition domain studied (low EVA contents), no significant evolution of glass transition, crystallization and melting temperatures was observed, corresponding evidently to immiscible blends. PP melting enthalpies remain identical as well, giving a crystallinity ratio almost constant over the composition range studied and equal to *ca.* 42 %.

Thermal characteristics were identical for drawn and non drawn samples. In particular, no influence of drawing on PP crystallinity ratio was observed.

Three relaxation processes, designated α_{EVA}, α_{PP} and β_{PP} in order of increasing temperature, can be observed on MRS spectra obtained from samples without drawing and uniaxially drawn. The temperatures associated to these relaxations for different EVA contents in the blends are given in Table 1. The transition noted β_{PP} is assigned to a PP crystalline phase relaxation. For drawn samples, this transition appears only as a shoulder on the tan δ vs T curve. The positions of α_{EVA} and α_{PP} relaxations, which are associated to the EVA and PP motions in the amorphous phases respectively, do not show any significant evolution with the EVA content in the blend. This is a confirmation of the immiscibility of these blends. It can be can noticed that the α_{EVA} relaxation is not influenced by the blend morphology whereas an increase in the temperature associated to the α_{PP} transition is observed in blends containing EVA fibers.

Table 1. MRS values for PP-EVA blends (pressed and drawn samples).

% EVA	pressed samples			drawn samples	
	Tα EVA (°C)	Tα PP (°C)	Tβ PP (°C)	Tα EVA (°C)	Tα PP (°C)
0		6.5	96.1		6.9
2.5	n.o.	6.2	98.3	n.o.	15
5	-27	5.6	98.8	n.o.	14.1
10	-32.7	5.1	91.4	-31.3	11.3
20	-26.5	6.7	96	-28.3	10.7
30	-27.4	5.8	91.7	-36.5	9.1
100	-18.7	3.3		-19.2	

n.o.: not observed

T taken at the maximun of Tan δ

These results can be explained by the reinforcement effect of EVA fibers which favours a decrease in molecular mobility and gives rise to a higher $T_{\alpha PP}$ for drawn samples than for pressed samples.

Results obtained with Dielectric Relaxation Spectroscopy (DRS) on the same blends are significantly different from those observed on MRS spectra. Fig.3 and 4 show the dielectric data (log ε" = f(T)) at frequencies between 0.6 and 39100 Hz for the PP-EVA 90/10 pressed and drawn blend respectively.

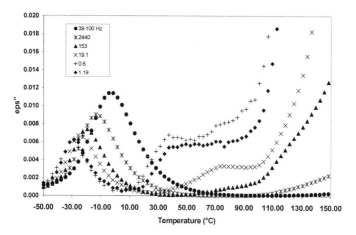

Figure 3. Dielectric loss as a function of temperature at several frequencies for PP-EVA 90/10 blend (pressed sample).

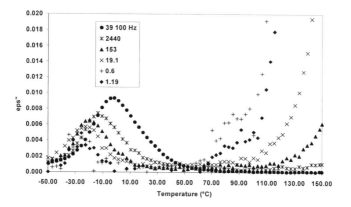

Figure 4. Dielectric loss as a function of temperature at several frequencies for PP-EVA 90/10 blend (drawn sample).

Two relaxation processes are detected for the pressed sample whereas only one relaxation process is observable for the drawn sample. The α_{EVA} relaxation, located at −28.7 °C at 1.19 Hz (pressed sample) and assigned to the EVA glass transition, shows only a slight dependence on the frequency and is observed at similar temperatures for both morphologies. The activation energy estimated for the α_{EVA} process is of the order of 275 kJ.mol^{-1} which is an expected value for such cooperative motions. The PP amorphous phase relaxation (α_{PP} transition) cannot be detected by DRS in these blends because of the low polarity of PP which does not give rise to sufficient fluctuations of dipolar moments to be measured efficiently.

An additional relaxation phenomenon can be noticed on the dielectric spectra of the pressed sample (observed at 52°C for 1.19 Hz) (Fig.3) which seems to be extremely frequency dependent. The activation energy associated to this relaxation can be estimated to 78.5 kJ.mol^{-1} which is characteristic of a weakly activated phenomenon. It could be attributed to an interfacial polarization effect corresponding to the probable migration of the free charges in the applied electrical field which are then blocked at the interface between the EVA droplets and the PP matrix. The values of temperature and frequency observed for this relaxation are in good agreement with theoretical calculations (presented in detail in a further article). Such a process cannot be observed for drawn samples in the range of frequency accessible in the present work. In fact, the modification of the EVA morphology directly influences the shape factor A (length of the long a to b short axis ratio for spheroids), resulting in too long relaxation times (too low frequencies) which are not experimentally measurable.

Conclusion

The dynamic and dielectric mechanical behaviours of PP-EVA blends have been reported as a function of morphology. Both DRS and MRS give evidence of the α-relaxation associated to the EVA glass transition without significant influence of EVA morphology. In contrast, a reinforcement effect is obtained with EVA fibers that contributes to a decrease in the mobility of PP, and therefore shifts the mechanical α_{PP}-relaxation to higher temperatures. An interfacial polarization effect is evidenced in some blends. Further investigations are in progress to obtain a better understanding of this phenomenon occurring in such heterogeneous systems in relation to the morphology of the dispersed phase.

Acknowledgements

We wish to thank P. Alcouffe for SEM observations and F. Melis for blend extrusion.

246

[1] E. Ramirez-Vargas, D. Navarro-Rodriguez, F.J. Medellin-Rodriguez, B.M. Huerta-Martinez and J.S. Lin, *Polym.Eng.Sc.* **2000**, 40(10), 2241; **2002**, 42(6), 1350

[2] R.L.McEvoy and S.Krause, *Macromolecules* **1996**, 29, 4258

[3] R.C.L.Dutra, B.G.Soares, M.M.Gorevola, J.L.G.Silva, V.L.Lorenco and G.E.Ferreira, *J. Appl.Polym.Sci.* **1997**, 66, 2243

[4] I.Ray and D.Khastgir, *Polymer* **1993**, 34 (10), 2030

[5] T.W.Smith, M.A.Abkowitz, G.C.Conway, D.J.Luca, J.M.Serpico and G.E.Wnek, *Macromolecules* **1996**, 29, 5042

[6] D.Hayward, R.A.Petrick and T.Siriwittayakorn, *Macromolecules* **1992**, 25, 1480

[7] B.Lestriez, A.Maazouz, J.F.Gerard, H.Sautereau, G.Boiteux, G.Seytre and D.E. Kranbuehl, *Polymer* **1998**, 39(26), 6733

[8] G.Perrier and A.Bergeret, *J.Appl.Phys.* **1995**, 77(6), 2651

[9] P.A.Aldrich, R.L.McGee, S.Yalvac, J.E.Bonekamp and S.W.Thurow, *J.Appl.Phys.* **1987**, 62(11), 4504

[10] A.Boersma and J.Van Turnhout, *J.Polym.Sci.* **1998**, 36, 2835

Crystalline Organization in Syndiotactic Polystyrene Gels and Aerogels

Christophe Daniel,* Gaetano Guerra

Dipartimento di Chimica, Università degli Studi di Salerno, Via S. Allende, 84081 Baronissi (SA), Italy
E-mail: cdaniel@unisa.it

Summary: The crystalline structure of syndiotactic polystyrene gels and aerogels has been investigated by using x-ray diffraction. Results show that, depending on the solvent, the crystalline structure of the junction zones of the gels is a clathrate phase or the solvent free orthorhombic β-form. For aerogels obtained from gels with a clathrate phase, the aerogel crystalline phase consists of the nanoporous δ-form while for aerogels obtained from gels with the β-form, the original crystalline structure is maintained.

Keywords: aerogels; crystalline structure; gels; syndiotactic polystyrene

Introduction

Syndiotactic polystyrene (sPS) displays a complex polymorphic behaviour and in the crystalline state four crystalline forms, designed by the acronyms α, β, γ, and δ, can be obtained.[1] The polymer chains adopt the all-trans planar zig-zag structure in the α– and the β–form while the s(2/1)2 helical conformation is present in the γ– and δ–forms. In addition to these four crystalline forms, semicrystalline clathrate structures characterized by the helical chain conformation can be obtained by sorption of suitable compounds (mainly halogenated or aromatic) in amorphous sPS samples as well as in sPS samples being in the α-, γ-, or δ-form.

It is well-known that sPS easily forms physical gels with several organic solvents and many reports focussing on the molecular structure, the morphology, the thermal behaviour and the crystalline structure have been published.[2] Depending on solvent-type and/or thermal treatments, the polymer-rich phase of the gels is characterized by the s(2/1)2 helical [2a,b,f,h] or the planar zigzag[2c,g,j] chain conformations.

© 2005 WILEY-VCH Verlag GmbH & KGaA, Weinheim   DOI: 10.1002/masy.200550432

It has been recently observed that complete removal of the solvent from sPS gels with the polymer-rich phase characterized by the s(2/1)2 helical or the planar zigzag chain conformations can be achieved by an extraction procedure based on supercritical carbon dioxide and high porosity physical aerogels can be obtained.[3]

In this paper we report on x-ray diffraction investigations relative to gels obtained with 1,2-dichloroethane (DCE) and 1,2-chlorotetradecane (CTD) and aerogels obtained from these different gel samples. In sPS/DCE gels polymer chains assume the s(2/1)2 helical conformation[2h] while in sPS/CTD gels polymer chains assume the planar zigzag chain conformation.[2j]

Experimental Part

The syndiotactic polystyrene used in this study was manufactured by Dow Chemicals under the trademark Questra 101. The ^{13}C nuclear magnetic resonance characterization showed that the content of syndiotactic triads was over 98%. 1,2-dicloroethane (DCE) and 1-chlorotetradecane (CTD) were purchased from Aldrich and used without further purification.

All sPS gel samples were prepared in hermetically sealed test tubes by heating the mixtures above the boiling point of the solvent until complete dissolution of the polymer and the appearance of a transparent and homogeneous solution had occurred. Then the hot solution was cooled down to room temperature where gelation occurred.

Aerogel samples were obtained by treating native gels with a SFX 200 supercritical carbon dioxide extractor (ISCO Inc.) using the following conditions: T= 45°C, P = 200 bar, extraction time t = 60 min (sPS/DCE gel), t = 180 min (sPS/CTD gel).

X-ray diffraction patterns were obtained on powder samples with nickel-filtered Cu K_α radiation with automatic diffractometers PW1710 (Phillips) and D8 (Bruker). Gel samples prepared beforehand in test tube and aerogels were reduced in fine powder before data collection was performed.

Results and discussion

In Figure 1, x-ray diffraction patterns of sPS/DCE gels prepared at different polymer concentrations are reported.

For polymer concentrations below 0.05 g/g, the amount of crystallites formed in gels is too small and diffraction patterns do not display any Bragg peak. By increasing the polymer concentration, the crystallinity of gel samples increases and for a gel prepared at 0.15 g/g weak peaks can be observed at $2\theta \approx 18$, 21, 22, and 29°. By increasing further the polymer concentration, diffraction peaks become sharper and more intense and for a gel prepared at 0.35 g/g, the diffraction pattern displays Bragg peaks located at $2\theta = 8.2$, 10.8, 18, 21, 24, and 29°. The location of the reflections which is substantially identical to that of sPS/DCE clathrate[2h,4] indicates that the cross-links domains of sPS/DCE gels is a crystalline clathrate phase.

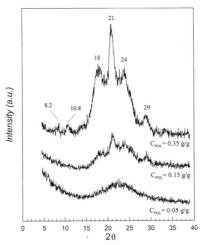

Figure 1. X-ray diffraction patterns for gels obtained in DCE. Concentrations are expressed as polymer weight fraction.

It is worth adding that similar x-ray diffraction patterns were obtained with sPS gels prepared with various solvents capable to form a clathrate phase (in particular chloropropane, toluene, decahydronaphthalene).

In Figure 2, x-ray diffraction patterns of sPS/CTD gels prepared at different polymer concentrations are reported.

Conversely to the gel obtained with DCE, we can observe diffraction peaks for a polymer concentration as low as 0.05 g/g (in particular at $2\theta \approx 12.5$ and 13.5°). This can be attributed to a higher gel crystallinity or a larger crystallite size. By increasing the polymer concentration as low as 0.05 g/g (in particular at $2\theta \approx 12.5$ and $13.5°$). This can be attributed to a higher gel crystallinity or a larger crystallite size. By increasing the polymer concentration, diffraction peaks become sharper and more intense and for a gel prepared at 0.20 g/g, the diffraction pattern displays Bragg peaks located at 2θ = 6.3, 10.5, 12.5, 13.7, 18.7, and 20.4 thus indicating the presence in the gel of the orthorhombic β-form.[5]

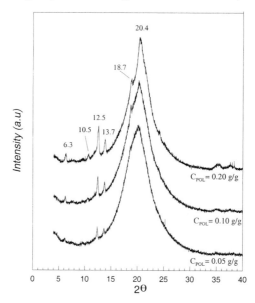

Figure 2. X-ray diffraction patterns for gels obtained in CTD. Concentrations are expressed as polymer weight fraction.

It has been recently observed that complete removal of the solvent from sPS gels can be achieved by an extraction procedure based on supercritical carbon dioxide and high porosity physical aerogels can be easily obtained.[3]

In Figure 3 are reported the x-ray diffraction patterns of aerogels obtained from gels prepared in DCE and in CTD at C_{POL} = 0.10 g/g.

The x-ray diffraction pattern of the aerogel obtained from the sPS/DCE gel displays strong reflections located at 2θ (CuKα) 8.3°, 13.5°, 16.8°, 20.7°, 23.5° and the absence of a diffraction peak at $2\theta \cong 10.6°$, thus indicating that the CO_2 treatment has extracted DCE also from the clathrate phase correspondingly producing a nanoporous δ–phase.[6] Aerogels with the nanoporous δ-phase were also obtained with gels prepared in solvent capable to form a sPS clathrate phase (in particular toluene, benzene, chloroform).

The diffraction pattern of the aerogel obtained from the sPS/CTD gel displays strong reflections at 2θ = 6.1°, 10.4°, 12.3°, 13.6°, 18.5° and 20.2° thus indicating the maintenance in the aerogel of the orthorhombic β-form, being already present in the starting gel.

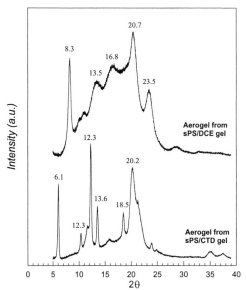

Figure 3. X-ray diffraction patterns for aerogels obtained from gels prepared in DCE and in CTD at $C_{POL} = 0.10$ g/g.

Conclusion

In this short communication x-ray diffraction chracterization of sPS gels prepared in DCE and CTD and of corresponding aerogels has been presented. In the former gels, the junction zones consist of a clathrate phase while for gels prepared with CTD the β-form is obtained. When the polymer rich phase of the physical gel is a clathrate phase, the corresponding aerogel is characterized by the nanoporous crystalline δ-phase while for aerogels obtained from the other type of gels the β-form is maintained.

It is worth adding that aerogels with the nanoporous δ-form are very promising for industrial applications on chemical separation and water purification as they present the high sorption capacity typical of s-PS δ–form samples associated with the high sorption kinetics typical of aerogels (due to the high porosity).[3]

Acknowledgements

Financial support of the "Ministero dell'Istruzione, del'Università e della Ricerca" (Prin 2002, Firb2001 and Cluster 26) and of Regione Campania (Legge 41 and Centro di Competenza) is gratefully acknowledged.

[1] (a) G. Guerra, V.M. Vitagliano, C. De Rosa, V. Petraccone, P. Corradini P., *Macromolecules* **1990**, *23*, 1539. (b) C. Manfredi, C. De Rosa, G. Guerra, M. Rapacciuolo, F. Auriemma, P. Corradini, *Macromol. Chem. Phys.* **1995**, 196, 2795.

[2] (a) M. Kobayashi, T. Nakaoki, N. Ishihara, *Macromolecules* **1990**, *23*, 78 (b) M. Kobayashi, T. Kosaza, *Appl. Spectrosc.* **1993**, *9*, 1417. (c) F. Deberdt, H. Berghmans, *Polymer* **1993**, *34*, 2192. (d) M. Kobayashi, T. Yoshioka, T. Kozasa, K. Tashiro, J.-I. Suzuki, S. Funahashi, Y. Izumi, *Macromolecules* **1994**, *27*, 1349. (e) M. Kobayashi, T. Yoshioka, M. Imai, Y. Itoh, *Macromolecules* **1995**, *28*, 7376.(f) C. Daniel, M.D. Deluca, J.M. Guenet, A. Brulet, A. Menelle, *Polymer* **1996**, *7*, 1273. (g) Y. Li, G. Xue, *Macromol. Rapid Commun.* **1998**, *19*, 549. (h) C. Daniel, G. Guerra, P. Musto, *Macromolecules* **2002**, *35*, 2243. (i) B. Ray, S. Said, A. Thierry, P. Marie, J.M. Guenet, Macromolecules **2002**, *35*, 9730. (j) C. Daniel, D. Alfano, G. Guerra, P. Musto, *Macromolecules* **2003**, *36*, 1713.

[3] G. Guerra, E. Reverchon, C. Daniel, V. Venditto, P. Mensitieri, Ital. Pat. 2003.

[4] C. De Rosa, P. Rizzo, O. Ruiz de Ballesteros, V. Petraccone, G. Guerra, *Polymer* **1999**, *40*, 2103.

[5] C. De Rosa, M. Rapacciuolo, G. Guerra, V. Petraccone, P. Corradini, *Polymer* **1992**, *33*, 1423.

[6] (a) C. De Rosa, G. Guerra, V. Petraccone, B. Pirozzi, *Macromolecules* **1997**, *30*, 4147. (b) G. Milano, V. Venditto, G. Guerra, L. Cavallo, P. Ciambelli, D. Sannino, *Chem. Mater.* **2001**, *13*, 1506.

Study of the Drying Behavior of Poly(vinyl alcohol) Aqueous Solution

Nadine Allanic, Patrick Salagnac,* Patrick Glouannec

Laboratoire d'Etudes Thermiques, Energétiques et Environnement, Centre de Recherche, BP 92116, 56 321 Lorient Cedex, France
E-mail: patrick.salagnac@univ-ubs.fr

Summary: This paper deals with the drying behavior of poly(vinyl alcohol) aqueous solution containing an active substance and placed into a Petri box. The objective is to reduce the drying time while respecting some constraints. To succeed, it is important to understand complex mechanisms governing heat and mass transfers. During the drying, the product thickness shrinks and its properties evolve. Drying kinetics in convective and infrared radiation are presented.

Keywords: diffusion; drying; poly(vinyl alcohol); thermal properties

Introduction

The aim of this work is to dry a polymer aqueous solution containing an active substance in a time compatible with an industrial production. Initially, the mixture with a 85% water volume fraction, is placed into a Petri box. The industrial drying process must be realised in a sterile atmosphere. As the proliferation of bacteria and the displacement of dust must be limited, convective drying with high air velocity is inadequate. Associating convective drying with infrared or microwave radiations is a solution frequently proposed in industrial process[1]. Indeed, the use of volatile organic solvents tends to be reduced and replaced by water. Due to the greater amount of energy necessary to evaporate water, mechanisms governing polymers drying have been investigated in lots of theoretical and experimental studies[2-4] to optimize energy requirement. Most of them have studied products with a low thickness. Lamaison[5] and Navarri[6] describe infrared drying, respectively of a painting and a PVA coating. Le Person[7] compares different drying modes (convection, conduction, infrared drying) of a thin pharmaceutical film. All these studies show that the direct energy supply appears to be an efficient solution. However, non-uniform radiant heating or high-intensity can damage polymer[8].

In our study, a short infrared drying, associated with a low air flow is chosen. Several constraints must be respected. The Petri box temperature must be lower than 90°C (Petri box deformation stresses). The final product must be a thin film with a precise moisture content and uniformly distributed at the bottom of the Petri box. During drying, the water evaporation from the surface creates concentration and temperature gradients. The thickness of polymer decreases, and its properties, including its microstructure, change considerably[9].

In the first section, in order to understand heat and mass transfers, a physical model taking into account film shrinkage has been established. Then, thanks to experiments, the properties of the product are determinated. Drying experimental curves in convection and infrared irradiation are presented. Finally, simulated and experimental results are compared for a convective drying.

Heat and mass transfers model

During the drying process, the effect of evaporation is a one-dimensional shrinkage along the normal at the air-product interface. Figure 1 gives heat and mass transfers and boundary conditions applied to the Petri box and the mixture. Inside the product, transfers are supposed one dimensional (x-axis). Due to the great thickness, temperature gradients are not negligible. The thickness can be expressed as a linear function of the average moisture content $\overline{X}$ [6,9]. Inside the product, mass transfer is liquid diffusion. The evaporation phenomena exists only on the air-product interface. The evaporation flux is given by[6]:

$$F_m = k_m \frac{M_v}{RT}\left(a_w P_{vsat} - P_{va}\right)$$ (1)

where k_m is the mass transfer coefficient, R the gaz constant, M_v the vapour molecular weight, P_{vsat} the saturated vapour pressure at surface of the mixture and P_{va} the air vapour pressure. Figure 2 gives activity a_w obtained by fitting experimental data by GAB model[10]. The curve has a similar profile to the Flory-Huggins model given for PVA by Perrin[11]. Activity is one of the key parameter of this study.

Thermal balance of the mixture and the conservation of the solvent are written as[5]:

$$\rho C_p \frac{\partial T}{\partial t} = \frac{\partial}{\partial x}\left(\lambda \frac{\partial T}{\partial x}\right) + \frac{\partial}{\partial x}\left(D_{AB}\left(C_{pA}\rho_A^0 - C_{pB}\rho_B^0\right)T \frac{\partial f_A}{\partial x}\right) + \phi$$ (2)

$$\frac{\partial f_A}{\partial t} = \frac{\partial}{\partial x}\left(D_{AB} \frac{\partial f_A}{\partial x}\right) \quad \text{with} \quad f_A = \frac{\rho^*}{\rho_A^0}\left(\frac{X}{1+X}\right)$$ (3)

where ρ, C_p and λ are respectively the density, the specific heat and the thermal conductivity of the mixture. D_{AB} is the diffusivity of water (A) diffusing through polymer (B) and f_A is the volume fraction. The infrared power, noted ϕ, is given by the Lambert-Beer law[12] as a function of the extinction coefficient κ and ϕ_0 the infrared radiation at the polymer surface:

$$\phi = \phi_0 e^{-\kappa x} \tag{4}$$

For the Petri box, thermal balance takes into account heat conduction and infrared radiation source term. The product and the Petri box are semi-transparent media. Hence, power must be correctly chosen in order to avoid skin formation at polymer surface and deformation of the Petri box.

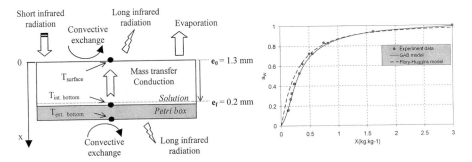

Figure 1. Heat and mass exchanges.　　　　Figure 2. Activity.

Results and discussions

Materials and experiments

The drying equipment, which has been described previously[13], enables us to combine convective and infrared drying. The infrared emitter, working in the short infrared range, is placed on the upper side of the chamber. Temperatures inside and outside the Petri box are measured by two thermocouples with a 75μm diameter (cf. Figure 1). An optical pyrometer gives the surface temperature of the polymer. An electronic scale is used to acquire the product mass as a function of time. Final products will be dry under a laminar flow in a sterile atmosphere. To have similar conditions to the industrial process, the air enters in the drying chamber with a low velocity ($v = 0.5$ m.s^{-1}) and a low relative humidity ($H_r = 20\%$).

Mixture diffusivity and infrared absorptivity determination

For low moisture content, diffusion is the main phenomena and determines drying kinetics. Therefore, diffusion coefficient is an important parameter to estimate. Usually, it is identified thanks to drying curves by inverse method[14]. An exponential dependence of the diffusion coefficient on the solvent content and temperature is proposed in literature for lots of polymer solutions[6,11,15]. The following expression has been chosen:

$$D_{AB} = D_0 e^{-\frac{E_{ad}}{RT}} e^{-\frac{a}{X}} \qquad (5)$$

The values of activation energy E_{ad} and coefficient a are the ones found by Navarri[6] for poly(vinyl alcohol) ($E_{ad} = 31700$ J.mol^{-1} and a = 0.332). In the case of constant diffusion coefficient and thickness, the Fick's second law has an analytical solution[15]. It allows to identify the initial diffusivity thanks to convective drying kinetics. Then, we deduce from this value, the coefficient $D_0 = 5.27 \ 10^{-6}$ m^2.s^{-1}.

In infrared drying, the evaporation rate and the temperature evolution depend on spectral distribution of the irradiation and radiative properties of the product. Figure 3 presents the polymer solution transmissivity (τ) for a thickness of 1.3 mm. These measures are obtained with a FTIR spectroscopy and compared with water spectrum[16]. The dimensionless spectral infrared irradiation is also presented. Therefore, the emitters are identified as a black body with an emissive temperature of 2000 K[12]. During drying, the equivalent transmittivity of the mixture is expressed as a function of the average water volume fraction and the thickness (e):

$$\tau = \tau_A \tau_B = e^{-\kappa_A \bar{f}_A e} e^{-\kappa_B (1-\bar{f}_A) e} \qquad (6)$$

where τ_A, τ_B and κ_A, κ_B are respectively the equivalent transmissivity and the total absorptivity of the water and the polymer. Thanks to latter curves, we can determinate $\kappa_A = 555$ m^{-1} and $\kappa_B = 2300$ m^{-1} and deduce equivalent transmittivity during drying. On the other hand, the equivalent transmissivity has been determined by measuring the total flux received by the product (Petri box + mixture) and transmitted. These measures were made with a sensor developed in the laboratory[17]. First, the total absorptivity of the Petri box was determined ($\kappa_{box} = 265$ m^{-1}). Figure 4 demonstrates that it exists a good agreement between the two methods. A small difference appears for low moisture content, maybe because we don't take into account reflexion at different interfaces. During drying, due to water evaporation, the product transmittivity varies. Changes in composition and a decreasing of the thickness involve a decreasing of absorptivity. At the end of the drying, the mixture absorbs only 15 % infrared irradiation. It is less than the irradiation absorbed by the Petri box.

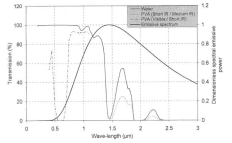

Figure 3. Transmissivity of water and PVA. and dimensionless spectral emissive power.

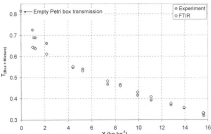

Figure 4. Evolution of the equivalent Transmissivity.

Drying kinetics analysis

The influence of infrared intensity on drying time and quality is now analyzed. Figure 5 gives variation of the evaporation rate as a function of moisture content for different initial values of the infrared intensity and for a time horizon of five minutes. Increasing infrared intensity enables us obtain a higher evaporation rate. However, the first drying stage (activity close to one) is smaller. The solvent concentration at the surface falls down quicker involving important decreasing of the evaporation rate and of the diffusion term. At this time, drying is no more controlled by evaporation, but limited by diffusion phenomena. Moreover, during this stage, surface temperature increases all the more so since infrared intensity is higher[6,18]. The polymer and the Petri box can be damaged, that is the reason why it is important to regulate energy inputs. In this aim, a model is developed, based on heat and mass equations previously presented (1-4). These are solved by control volume method with implicit scheme (number of control volumes: 50, time step: 0.2 s). Figure 6 presents experimental and simulated results for a convective drying. Until 6000 s, a good agreement is observed between simulated and experimental curves. After this time, diffusivity is certainly surrounding, involving higher temperatures than experimental ones. The experimental evaporation rate decreasing is also not important enough, maybe because of skin formation at the surface or changes in microstructure[17]. This test shows that it is necessary to estimate a better diffusion coefficient using inverse methods before trying to simulate infrared drying kinetics.

258

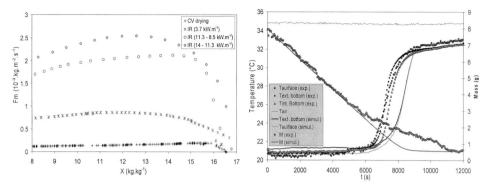

Figure 5. Evaporation rate evolution vs moisture content for different initial powers.

Figure 6. Convective drying kinetics.

Conclusion

In this study, mechanisms limiting drying of poly(vinyl alcohol) have been analyzed. Absorptivity and diffusivity, which are key parameters of this problem, are estimated as a function of moisture content and of other parameters (temperature, thickness,...). Experimental drying kinetics have shown the importance of the energy input adapted to the mixture and Petri box thermal behavior. To optimize drying time, it is necessary to anticipate skin formation and to find a way to delay it.

[1] Jonkers G., Nienhuis J.G., Van De Velde B., *Macromolecular Symposium*, **2002**, 187, 249-260.
[2] Ventras J.S., Ventras M., *Journal of Polymer Science, Part B : Polymer physics*, **1994**, 32, 187-194.
[3] Yoshida M., Miyashita H., *Chemical engineering Journal*, **2002**, 86, 193-198.
[4] Carra S., Pinoci D., Carra S., *Macromolecular Symposium*, **2002**, 187, 585-596.
[5] Lamaison V., Scudeller Y. & al, *International Journal of Thermal Science*, **2001**, 40, 181-194.
[6] Navarri P., Andrieu J., *Chemical Engineering and Processing*, **1993**, 32, 319-325.
[7] Le Person S., Puiggali J.R. & al., *Chemical Engineering and Processing*, **1998**, 37, 257-263.
[8] Chen J.J., Lin J.D., *International Journal of Heat and Mass Transfer*, **2000**, 43, 2155-2175.
[9] Guerrier B., Bouchard C., Allain C., Bénard C., *AIChE Journal*, **1998**, 44, 4, 791-798.
[10] Srinivasa P.C., Ramesh M.N. & al., *Carbohydrate Polymers*, **2003**, 53, 431-438.
[11] Perrin L., Nguyen Q.T., Clement R., Noel J., *Polymer International*, **1996**, 39, 251-260.
[12] Incropera F.P., De Witt D. P., *Fundamentals of heat and mass transfer*, 5th ed., Eds. Wiley & Sons, New York 2002.
[13] Glouannec P., Lecharpentier D., Noel H., *Applied Thermal Engineering*, **2002**, 22, 1689-1703.
[14] Doumenc F., Guerrier B., *AIChE Journal*, **2001**, 47, 5, 984-993.
[15] Ion L., Vergnaud J.M., *Polymer Testing*, **1995**, 14, 479-487.
[16] Hale G. M., Querry M. R., *Applied Optical*, **1973**, 12, 555-563.
[17] Noël H., Ploteau J.P., Glouannec P., *Proceedings, 8th International Symposium on Temperature and Thermal Measurements in Industry Science*, **2001**, 925-930.
[18] Yoshida M., Miyashita H., *Chemical Engineering Journal*, **2002**, 86, 193-198.

Macromol. Symp. **2005**, *222*, 259-263

Relations between Glass Transition Temperatures in Miscible Polymer Blends and Composition: From Volume to Mass Fractions

Sylvie Pimbert, Laurence Avignon-Poquillon,* Guy Levesque

Laboratoire Polymères, Propriétés aux Interfaces et Composites (L2PIC), Université de Bretagne-Sud, Centre de Recherche, rue de St Maudé, 56325 Lorient, France
E-mail: laurence.avignon@univ-ubs.fr

Summary: The use of volume fractions in the empirical mixing laws to predict the glass transition temperatures (Tg) of polymer blends provides good agreement with experimental values, even for polymer systems with different densities. No adjustment parameter is therefore required whereas Gordon-Taylor and Kwei equations based on weight fractions need the use of a fitting parameter which has to be determined from experimental data. This assumption was validated from Tg measurements through DSC experiments conducted on PMMA /PVDF blends which have significantly different densities.

Keywords: empirical relations; glass transition; polymer blends

Introduction

A large amount of theoretical and practical work has been carried out in an attempt to correlate the glass transition temperature to composition in miscible polymer blends [1-7]. Some relations previously developed for copolymer glass transition temperature prediction have found applications to this purpose. However most practical correlations are based on weight fractions whereas theoretical ones use volume fractions. Obviously no significant difference occurs between these correlations if specific volumes or densities of both polymers are very similar. But such an approximation induces large deviations if the component mixture presents quite different densities. This paper illustrates how the use of volume fractions affords straightforward correlations between Tg and composition in blends containing polymers having largely different densities such as poly(methyl methacrylate) (PMMA) and semi-crystalline vinylidene difluoride (VDF) homopolymers.[1] The main difficulty in using volume fractions lies in the determination of the amorphous phase density in semi-crystalline polymers. For several blends, this parameter could be either calculated from literature and/or experimental data or estimated to be sufficiently near the semi-

          DOI: 10.1002/masy.200550434

crystalline polymer density if the crystalline fraction remains low enough. That is obviously the case in most VDF copolymers.

Experimental

Fluorinated polymers are commercially samples, available in pellet form: PVDF (SOLEF 1008), from Solvay (Belgium). PMMA OF 104S was supplied by Röhm. Blends were obtained by solution mixing followed by fast precipitation. Both polymers were dissolved at room temperature in dry dimethylformamide as 5% (w/w) solutions and aliquots mixed in the required amounts. After 24h stirring, the blended solutions were poured into a 100-fold excess volume of distilled water under stirring. The blends were filtered off, washed several times with water and dried at 70°C to constant weight. IR-spectra were recorded to check the absence of any residual solvent.

A Perkin–Elmer differential scanning calorimeter Pyris 1 was used to determine the transition temperatures. All samples were submitted to the same temperature program for anisotherm experiments: first heating from 12 to 210°C at 20°C min^{-1}, followed by cooling to 12°C at 10°C min^{-1}; the second scan was performed at 10°C min^{-1}.

Discussion

Several empirical relations have been proposed to correlate the glass transition temperature to miscible blend compositions: the most frequently used are the Fox [2] and Gordon-Taylor [3] equations which correlate the glass transition temperature Tg of the blend to the pure component Tg$_1$ and Tg$_2$ through their weight fractions W$_1$ and W$_2$ in the mixture.

Kwei et al.[4] have introduced into the Gordon-Taylor equation a corrective term (qW$_1$W$_2$) to quantify specific interactions between the two polymers :

$$Tg = \frac{Tg_1 W_1 + kTg_2 W_2}{W_1 + kW_2} + qW_1 W_2 \qquad (1)$$

Couchman[7] has developed theoretical relations that afford prediction of glass-transition temperature in miscible polymer blends. One of the simplest form for these relations is

$$Tg = \phi_1 . Tg_1 + \phi_2 . Tg_2 + \phi_1 \phi_2 . \Delta w / \Delta C_p. \qquad (2)$$

in which Δw represents the difference between the mixing pair interaction energy variation in both the liquid and glassy states and ΔC_p the usual heat capacity difference. Lu and al.[8] have demonstrated a relationship between the glass transition temperature and the interaction parameter of miscible binary polymer blends but they transform a term containing mole fractions into weight fractions.

Empirical relations use weight fractions whereas Couchman's relations are based on volume fractions. In most published papers concerned with glass transition temperature in miscible polymer blends, weight fractions are the unique parameters to be considered for practical reasons. If volume fractions are to be used, amorphous phase densities of both polymers are required to calculate volume fractions in the blended material : such data are not usually available from simple density measurements on semi-crystalline polymers. Crystallographic data afford a way to calculate the crystalline phase density and then to evaluate the amorphous phase density if the crystalline fraction may be determined (either from X ray diffraction or melting enthalpy measurements). This is the case for PVDF homopolymer. For copolymers, including VDF-copolymers, the crystallinity is obviously lower than for homopolymers. Then a rough approximation lies in the use of bulk material density instead of the amorphous phase density, assuming the cristallinity to be quite low. Such an approximation induces an error in density which remains $\leq 5\%$ in VDF copolymers having cristallinity in the range of 30-50% of the PVDF homopolymer cristallinity. Such an approximation seems to be lower than the deviations induced by the use of weight fraction in empirical relations.

But Tg evolution might be studied by another simple relation : a "mixing law" with volume fractions.

$$Tg = \phi_1.Tg_1 + \phi_2.Tg_2 \qquad\qquad (3)$$

with ϕ_1 and ϕ_2 the volume fraction of component 1 and 2 and Tg_1 and Tg_2 the glass transition temperature of the same components.

The volume fractions can be written as
$$\Phi_1 = \frac{\dfrac{W_1}{\rho_1}}{\dfrac{W_1}{\rho_1} + \dfrac{W_2}{\rho_2}}$$

(ρ_i are amorphous polymer densities)

Thus equation 3 reduces to :

$$Tg = \frac{W_1 Tg_1 + k W_2 Tg_2}{W_1 + k W_2} \qquad \text{with} \qquad k = \frac{\rho_1}{\rho_2}$$

which is « Gordon-Taylor » type relation with a constant k defined as the ratio between polymer densities in the amorphous state.

We have tested this simple « volume mixing law » in systems where the density ratio is far from unity so that in such blends volume fractions are quite different from weight fractions.

Moreover we have chosen systems in which the amorphous specific mass (or density) could be calculated from literature data or evaluated through reasonable approximations.

PVDF Containing Systems

Polyvinylidene difluoride (PVDF) has been extensively used in polymer blend miscibility studies : its well known crystallographic data and density measurements allow to use accurate amorphous density values ($\rho_a = 1.68$ g/cm^3). The density of PMMA is $\rho = 1.19$g/cm^3. In the PMMA/ PVDF blends, neither Gordon-Taylor, nor Fox equation apply whereas the Kwei equation (with q = 82.7) offers a good correlation between experimental results and theoretical values (fig. 1). However a similar good fit is also observed using the simple mixing law with the volume fractions without any adjustable parameter (fig. 2). The divergence occurs whatever the equation used when the PVDF content is higher than 50% because the cristallinity becomes important. The crystalline phase yields phase separation but the miscible amorphous phase remains with a unique constant Tg.

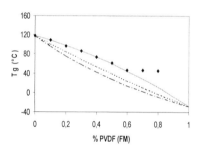 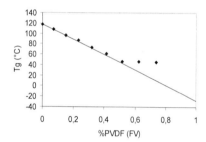

Figure 1. Tg evolution in PMMA-PVDF blends Experimental data (points); Kwei law (solid curve); Gordon-Taylor law (doted curve); Fox law (dash doted curve).

Figure 2. Tg evolution in PMMA-PVDF blend correlation with simple volume fraction mixing law (Solid curve).

VDF Copolymers Containing Systems

Other VDF copolymers were investigated and the experimental Tg values fit also very well with the simple « volume mixing law » postulated as a first order approximation of Couchman calculations. In some cases small deviations are noted for compositions having volume fractions $\phi > 0,30$ in semi-crystalline polymers. This is ascribed to the limit of miscibility imposed by the cristallinity induced phase separation. This must be compared to PVDF-

PMMA blends for which analysis through the classical Fox or Gordon-Taylor relations (figure 1) yields more important deviations even at $\phi = 0,10$ (lower values were not tested). This approach using volume fractions could therefore be applied to other semi-crystalline polymers or copolymers for which the density of the amorphous phase is accessible by calculation or measurement.

Conclusion

Experimental Tg's obtained through measurements on PMMA/PVDF or VDF copolymers blends have been compared to several empirical models used to predict polymer blend glass transition temperatures. The use of volume fractions instead of weight fractions in the empirical mixing equations provides good agreement with experimental values even for polymers with quite different densities. No adjustable parameter is required whereas in Gordon-Taylor and Kwei equations based on weight fractions a fitting parameter is necessary.

[1] Jouannet, D; Pham, T.N.; Pimbert,S.; Levesque, G.; *Polymer, 1997, 38, 5137*
[2] Fox, T.G; *Bull. Am. Phys. Soc.,1956, 1, 123*
[3] Gordon, M.; Taylor, J.S.; *J. Appl. Chem., 1952, 2, 493*
[4] Kwei, T.K.; *J. Polym. Lett. Ed., 1984, 22, 307*
[5] Brekner, M. J.; Schneider, H. A.; Cantow, H. J., *Polymer, 1988, 78*
[6] Schneider, H.A., *J. Therm. Anal. Cal., 1999, 56, 983*
[7] Couchman, P.R., *Macromolecules, 1978, 11, 1156*

Characterization of Interdiffusion between PVDF and Stereoregular PMMA by Using ATR-FTIR Spectroscopy

Gildas Lorec,* Christophe Baley, Olivier Sire, Yves Grohens

Laboratoire Polymères, Propriétés aux Interfaces et Composites (L2PIC) Rue de Saint-Maudé, 56321 Lorient Cedex, France
Fax (+33) 2 97 87 45 88; E-mail:gildas.lorec@univ-ubs.fr

Summary: In this study we investigated the interdiffusion between PVDF and PMMA below the melting temperature of PVDF by ATR-FTIR spectroscopy. The influence of the stereoregularity of different PMMA samples was studied. The PMMA tacticity showed a significant influence on the kinetics of the diffusion. Syndiotactic PMMA diffuses faster than isotactic and atactic PMMA which can be explained from the difference of chains stiffness between the two stereoisomers.

Keywords: diffusion; dynamic; PMMA; PVDF; tacticity

Introduction

The problem of the polymer-polymer interfaces is crucial because it plays a critical role in many fields like welding, adhesion, coextrusion and polymers mixtures. The characterization of the interface is essential because it is in many applications the essential feature of the final material mechanical properties. Interdiffusion phenomenon is strongly connected to the formation of the interphases in polymers. Interdiffusion between polymers is closely related to temperature, chemical composition, compatibility between polymers, molecular weight, distribution of the molecular weight, chains orientation and polymers microstructure.

Three classes of diffusion behaviour have been distinguished : Case I or Fickian diffusion, Case II diffusion and non-Fickian or anomalous diffusion. Experimentally, the type of diffusion can be established by the observed time dependence, t^n, where n is a constant. Case I systems are characterized by an exponent $n=\frac{1}{2}$, Case II by $n=1$ and non-Fickian systems by n taking an intermediate value between $\frac{1}{2}$ and 1.

Various therories have been proposed for the diffusion of polymer in melts. The most widely accepted theory is the model of reptation proposed by de Gennes[1] and Doi and Edwards[2]. The basis of reptation is the snakelike motion of a polymer chain along its own contour

© 2005 WILEY-VCH Verlag GmbH & KGaA, Weinheim
DOI: 10.1002/masy.200550435

formed by the constraint of neighbouring chains. The reptation model identifies distinct regions, which exhibit characteristic power-law time dependencies with exponents (n) equal to ¼ and ½. Interdiffusion between many polymer couples has been studied for several years[3-6].

Many studies of Poly(vinylidene fluoride) (PVDF)/poly(methyl methacrylate) (PMMA) blends have been carried out from both scientific and technological view points[7-14].

Several studies have been carried out on blends of PVDF and isotactic-, syndiotactic and atactic-PMMAs. Roerdink and Challa[9] reported the influence of tacticity of PMMA on the compatibility with PVDF based on observations of glass transition temperature (T_g) and melting temperature (T_m) depressions. Riedl and Prud'homme[10] evaluated the thermodynamic interaction parameter (χ_{12}) for PVDF/PMMAs with different tacticity using T_m depression data[9]. Both studies suggest that blends of PVDF and isotactic-, syndiotactic- and atactic-PMMAs are miscible. The lower value of the interaction parameter observed for the PVDF/i-PMMA system seems to indicate that the interaction of PVDF segments with i-PMMA segments is stronger than that with a-PMMA and s-PMMA segments. However, the values of χ_{12} for PMMA/PVDF blends estimated from the equilibrium melting temperature ($T°_m$) depression increase in the order atactic, syndiotactic and isotactic[15]. The χ_{12} value (-0.02) of PVDF/iso-PMMA suggests that the mixing state of PVDF/i-PMMA in the melt is metastable[15] which differs from the previous results. This could be due to the fact that χ_{12} values obtained by melting temperature depression using the Nishi-Wang equation yield large errors compared to values obtained by other methods[16]. Sasaki and al. have studied miscibility of PVDF/PMMA blends by crystallization dynamics[17]. The tacticity difference between at-PMMA and s-PMMA used in their study was not so large, however, both blends showed different crystallization dynamics. These results indicate that a slight difference in tacticity influences the miscibility of the blends. In contrast with previously reported results, they conclued that PVDF/i-PMMA were immiscible.

Benedetti and al. [18] investigated PVDF/PMMA blends by FTIR microspectroscopy and DSC. On the basis of the major shift of the carbonyl band of i-PMMA in the mixtures, they have observed stronger interactions for PVDF/i-PMMA compared with PVDF/s-PMMA. Strong interaction of PVDF with stereoregular PMMA have also been reported by Roerdink and Challa by infra-red spectroscopy[19].

The purpose of this study is to investigate the molecular interdiffusion across a PMMA/PVDF interface by using FTIR-ATR spectroscopy. These two polymers are known to be miscible in the molten state[7] but actually interdiffusion between these two polymers has not been observed below the melting temperature of PVDF. This is not trivial since PVDF high cristallinity (about 60%) is a limiting factor for interdiffusion.

In this paper, the influence of tacticity of PMMA on interdiffusion process will be described, espcially concerning the composition of the interface, in order to compare it with classical PMMA/PVDF blends. Analysis of selected infrared specific absorption bands permitted the tracking of diffusion of both components.

Experimental

The infrared spectra were obtained on a Perkin Elmer spectrometer. Spectra were collected at a resolution of 4 cm[-1] with 32 averaged scans. An electrically heated 6 reflections trough ATR cell bought from Specac LTD. was used. The cell can be heated up to 200°C. The temperature of the sample is measured with a thermocouple and regulated to ± 0.5°C with a Specac 3000 Series[TM] RS232 controller. The internal reflection element crystal was made from zinc selenide (ZnSe), having a refractive index of 2.42. The angle of incidence was chosen as 45°C. The extinction coefficients $[\varepsilon_i(\mu m^{-1})]$ of the PMMA C=O stretching mode at 1724 cm[-1] and PVDF C-F bending mode at 1402 cm[-1] were determined as the slopes of plots of absorbance against film thickness from the transmission spectra of a series of DMF or toluene-cast homopolymer films. Least squares slopes give $\varepsilon_{PVDF}= 0.1562 \mu m^{-1}$ and $\varepsilon_{PMMA}= 0.105 \mu m^{-1}$.

Poly(metylmethacrylate) (PMMA) was supplied by Polymer Source, Inc.. The Poly(vinylidene fluoride) (PVDF) was provided by Solvay. The main characteristics and origin of the polymers used in this study are shown in table 1.

The thin film of PVDF was cast directly from solution onto the ATR crystal, in order to achieve a good contact between the polymer and the ATR crystal. Since PVDF was soluble in dimethylformamide (DMF), a small amount of PVDF was dissolved into DMF at a concentration of 4,5 gram per liter. The system was stirred for 2h at 100°C to ensure complete mixing. 1 mL of the solution was cast in the ATR cell and then the film was dried at 80°C under vacuum for 12 hours. The thickness of the PVDF film was typically on the order of 1 μm which corresponds to the penetration depth[22], dp, of the infrared ray for the carbonyl pic (1724 cm[-1]). A thick film of PMMA of about 0.5 mm thick was pressed against the PVDF film under a pressure of 10 MPa.

Table 1. Main characteristics of polymers.

Polymer	Source	M_w (g.mol^{-1})	M_w/M_n	T_m (°C)	T_g (°C)
PVDF	Solvay	125000	1.2	174	-32
at-PMMA-76	Atofina	76000	2.08	-	100
at-PMMA-48,5	Polymer Source	48500	1.66	-	121
iso-PMMA-46	Polymer Source	46000	1.27	-	59
syn-PMMA-37,5	Polymer Source	37500	1.25	-	129

Results and Discussion

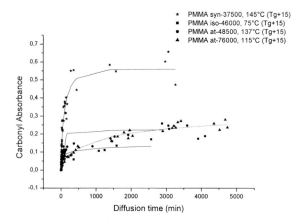

Figure 1. Plot of carbonyl absorbance versus diffusion time for PMMAs with different tacticities at T=Tg +15°C.

The evolution of the carbonyl absorbance with the diffusion time at T_{gPMMA}+15°C is shown in Figure 1. The ATR-FTIR measurements reveal a significant diffusion of PMMA in PVDF. Figure 1 shows that PMMA tacticity influences the diffusion. The absorbance of syndiotactic PMMA reaches 0.55 at the plateau at T_g+15°C (145°C) whereas absorbances of atactic and isotactic PMMAs are about 0.18. We assume that the chain mobility in the amorphous phase of PVDF is constant within the temperature range of 75°C to 145°C. T_g of PVDF is about – 32°C so the difference of temperature between all the experiments is very low compared to the difference with the T_g of PVDF. Experiments were also carried out at T=T_{gPMMA}+20°C and we observe the same differences of diffusion with the tacticity as it is shown on Figure 2.

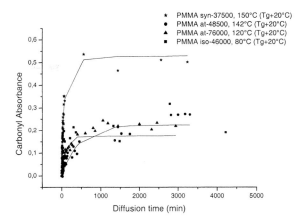

Figure 2. Plots of carbonyl absorbance versus diffusion time for PMMAs with different tacticities at T=Tg +20°C.

With ε_i in μm^{-1}, the measured apparent surface volume fraction of PMMA (ϕ_{PMMA}) within the sampling depth of the ATR technique is given by[21] :

$$\phi_{PMMA}=C_{PMMA}/(C_{PMMA}+C_{PVDF})\approx A_{PMMA}/(A_{PMMA}+\lambda A_{PVDF})$$

where $\lambda=dp_{PMMA}\varepsilon_{PMMA}/dp_{PVDF}\varepsilon_{PVDF}$. The respective depths of penetration are calculated according to Harrick[22] at 1724 cm^{-1} for PMMA and at 1402 cm^{-1} for PVDF. With n_1=2.4 and n_2=1.42 (the refractive index of PVDF), dp_{PMMA}= 0.993 μm and dp_{PVDF}= 1.22 μm. Values of ε_i were found a little higher than reported in the literature [21].

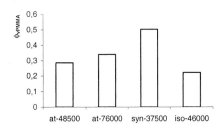

Figure 3. ϕ_v of PMMA in PMMA/PVDF interface after diffusion at T=T$_{gPMMA}$+15°C.

Figure 3 shows the evolution of the apparent surface volume fraction of PMMA in the measurement depth of the infrared evanescent beam. For syndiotactic PMMA, approximately 50% in volume of PMMA in the probed region is reached after two days of diffusion whereas only 30% is reached for atactic PMMA and 20% for isotactic PMMA. If we consider that the miscibility of the two polymers is the only parameter which influences the diffusion, these results would indicate that the interaction of syndiotactic PMMA segments with PVDF is stronger than the interaction with isotactic and atactic PMMAs. This is not in agreement with the literature data concerning the calculated Flory-Huggins interaction parameter[9]. High cristallinity of PVDF reduces considerably the mobility of the amorphous phase and the PMMA is only able to diffuse in 41% of the PVDF film. We can conclude from these results that the diffusion of PMMA chains in PVDF amorphous phase is more directly driven by dynamic phenomenons than by blend thermodynamics.

An important property that needs to be mentioned is the critical molecular weight for entanglements, M_c. Chain entanglement, a key concept in the dynamics of polymer chains in the melts, is affected by the polymer chain's flexibility. Chain entanglement, therefore, varies with chain tacticity in PMMA. In other words, M_e, hence M_c decreases with increasing syndiotacticity of the polymer samples. This reveals that PMMA actually has a range of M_c from about 11,650 for highly syndiotactic to about 40,000 for purely isotactic polymer[23]. Regarding these values, the reptation time of the isotactic chains might be lower than the one for syndiotactic chains which is obviously not the case[24]. Even though we consider the slight molecular weight difference of our samples, the reptation model is not consistent with the previous observations. Nevertheless, at the plateau where an equilibrium is reached, the large stiffness of isotactic PMMA[25] may also partially hinder its diffusion between the PVDF cristallites. The tortuous pathway of the amorphous regions of PVDF may prevent diffusion of stiff chains.

Conclusion

In this study, diffusion of PMMA chains in PVDF has been shown below the melting temperature of PVDF. Despite of the high cristallinity of PVDF, PMMA diffuses into amorphous phase of PVDF. The tacticity of PMMA has a significant influence on the diffusion which is not in agreement with the interaction parameters found in the literature. The stiffness of the PMMA chains is suggested to be relevant to the diffusion process. Further investigations are in progress to confirm the relevance of this parameter in polymer diffusion processes.

[1] P. G. de Gennes, *J . Chem. Phys.* **1971**, *55*, 572.

[2] M. Doi and S. F. Edwards, "The Theory of Polymer Dynamics", Oxford University Press, New-York 1986.

[3] E. Jabbari and N. A. Peppas, *Macromolecules* **1993**, *26*, 2175.

[4] C. M. Laot, E. Marand and H.T. Oyama, *Polymer* **1999**, *40*, 1095.

[5] R. Neuber and H.A. Schneider, *Polymer* **2001**, 42, 8085.

[6] J. G. Van Alsten and S. R. Lustig, *Macromolecules* **1992**, *25*, 5069.

[7] D. R. Paul and J. O. Altamirano, *Adv. Chem. Ser.* **1975**, *145*, 371.

[8] C. Leonard, J. L. Halary and L. Monnerie, *Polymer* **1985**, *26*, 1507.

[9] E. Roerdink and G. Challa, *Polymer* **1978**, *19*, 173.

[10] B. Riedl and R. E. Prud'homme, *Polym. Eng. Sci.* **1984**, *24*, 1291.

[11] C. Leonard, J. L. Halary and L. Monnerie, *Macromolecules* **1988**, *21*, 2988.

[12] W. Kaufmann, J. Petermann, N. Reynole, E. L. Thomas and S. L. Hsu, *Polymer* **1989**, *30*, 2147.

[13] S. Pimbert, L. Avignon-Poquillon and G. Levesque, *Polymer* **2002**, *43*, 3295.

[14] G. A. Gallagher, R. Jakeways and I. M. Ward, *J. Polym. Sci. Part B : Polym. Phys.* **1991**, *29*, 1147.

[15] H. Takimoto, Y. Sato, H. Yoshida, E. Ito and T. Hatakeyama, *Polym. Prepr. Jpn* **1993**, *42*, 1262.

[16] M. Takahashi, J. Hasegawa, S. Shimono and H. Matsuda, *Netsu sokutei* **1995**, *22*, 2.

[17] H. Sasaki, P. K. Bala, H. Yoshida and E. Ito, *Polymer* **1995**, *36*, 4805.

[18] E. Benedetti, S. Catanorchi, A. D'Alessio, P. Vergamini, F. Ciardelli and M. Pracella M, *Polymer International* **1998**, *45*, 373.

[19] E. Roerdink and G. Challa, *Polymer* **1980**, *21*, 509.

[20] W. E. J. R. Maas, C. H. M. Papavoine, W. S. Veeman, *J. Polym. Sci. Part B : Polym. Phys.* **1994**, *32*, 785.

[21] J. M. G. Cowie, B. G. Devlin and I. J. Mc Ewen, *Polymer* **1993**, *34*, 501.

[22] N. J. Harrick, "Internal Reflection Spectroscopy", Wiley Interscience, NewYork 1967.

[23] M. M. Hassan, C. J. Durning, *J. Polym. Sci. Part B : Polym. Phys.* **1999**, *37*, 3159.

[24] P. G. de Gennes, "Scaling Concepts in Polymer Physics", Cornell University Press, New York 1979.

[25] M. Vacatello, P. Flory, *Macromolecules* **1986**, *9*, 405.

Macromol. Symp. **2005**, *222*, 273-280

Smart Poly(styrene)/Carbon Black Conductive Polymer Composites Films for Styrene Vapour Sensing

Jean-François Feller, * *Hervé Guézénoc, Hervé Bellégou, Yves Grohens*

Laboratory of Polymers, Properties at interfaces & Composites, University of South Brittany, Saint-Maudé street, F56321 Lorient, France
E-mail: jean-françois.feller@univ-ubs.fr

Summary: Conductive Polymer Composites (CPC) can be used to elaborate sensing elements able to detect solvent vapours at very low concentrations (some ppm). Our experiments have shown that combining atactic PS or syndiotactic PS to five carbon black of different specific surfaces, allows obtaining a wide range of electrical resistances and surface morphologies. The CPC films have been elaborated from solutions by spraying and spin coating, the former being more adequate to design sensitive films with tuneable electrical properties. The larger electrical responses were obtained with an initial resistance close to 10^4 Ω. Our sensors gave a response for very low styrene concentration (some ppm) increasing as a function of vapour concentration.

Keywords: carbon black; conductive polymer composites; poly(styrene); vapour sensing

Introduction

Polymer films are often used in sensors design [1-3]. But in particular, associating an insulating polymer matrix to a conductive filler which provides electrically conductive polymer composites (CPC) leads to materials with very interesting sensing properties. In fact, the important resistivity variation of CPC with thermal [4], mechanical [5] or chemical [6-10] solicitations make possible their use as transducers. For vapour sensing, expected characteristics are: large electrical response, short response time, high sensitivity and important vapour selectivity. The first point is related to the CPC initial resistivity, the second to the CPC thickness, the third to the CPC specific surface and the later to the interactions between solvent vapour molecules and CPC matrix macromolecules. The design of low cost sensors able to detect volatile organic compound (VOC) such as methanol, toluene, chloroform and styrene used in composites processing or chemistry industries is of interest with regard to the new environment standards being applied in Europe.

In this study we have investigated the influence of several parameters as the CPC deposition process, the initial resistivity and the vapour concentration, on some poly(styrene)/carbon black (PS/CB) sensors electrical response to styrene vapours.

DOI: 10.1002/masy.200550436

Experimental

Materials: Five carbon blacks (CB) of different specific surface areas (cf. **Table 1**) have been dispersed in two poly(styrene) (PS) matrices atactic and syndiotactic (cf. **Table 2**) in solution. Toluene, o-dichlorobenzene and styrene were provided by ACCROS.

Table 1. Polymers characteristics.

	sPS	aPS
T_g (°C)	98	105
T_m(°C)	270	-
$T_{c,n}$(°C)	242	-
M_n (g.mol^{-1})	94 100	2 500
M_w (g.mol^{-1})	192 000	-
Tacticity index (%)	0.99	-
$\Delta H_m/\Delta H_\infty$ (J.g^{-1})/ (J.g^{-1})	53.2/98	-
Density (g.cm^{-3})	1.05	1.05
Supplier	Resinex- Dow	PolySciences

Table 2. Carbon blacks characteristics.

Supplier	Degussa	Degussa	Erachem	Erachem	Erachem
Type	Corax N550	N115Corax	Ensaco150G	Ensaco250G	Ensaco350G
BET N_2 (m^2.g^{-1})	44	143	50	65	770
CTAB (m^2.g^{-1})	42	128	-	-	-
DBP (cm^3.100g^{-1})	121	113	165	190	320
Diameter (nm)	40	29	40	38	35

Sample preparation: aPS flakes were first introduced in the reactor and then dissolved in toluene under stirring at 100°C for 1 hour to obtain solutions at 20 g.dm^{-3}. sPS pellets, less soluble due to their high melting temperature (T_m) and crystallinity (X%) were dissolved in o-dichlorobenzene at 150°C for 2 hours at a lower content 5.5 g.dm^{-3} to prevent gel formation. In a second step, carbon black was introduced in the reactor under sonication for 1 hour to obtain homogeneous solutions with a CB/PS weight ratio of about 12/100.

Sensor processing: Two deposition processes where used, spraying and spin coating. For the first process an air pressure of 0.2 MPa was applied for 1.30 s at respectively 100°C and 150°C for aPS and sPS for each layer deposition whereas for the second process, drops of 10^{-3} cm^3 were deposited under rotation (index 2) during 10 s. The sensor substrate consisted in interdigitated copper electrodes engraved in a typical epoxy board (cf. **Figure 1**). After spraying, the surface of the substrate is coated with a homogeneous CPC film (cf. **Figure 2**).

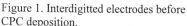

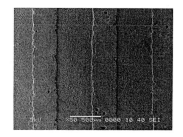

Figure 1. Interdigitted electrodes before CPC deposition.

Figure 2. SEM picture of the CPC sensor.

Characterization: *Electrical resistivity* was derived from the intensity and voltage measurement with a KEITHLEY 2000 multimeter. Tension was adjusted with a stabilized source. Considering the important resistance of the samples, a two-probe technique was used. Collection and processing of data was done by an acquisition program developed with VISUAL DESIGNER 4.0 described in a previous work [9]. *Calorimetric measurements* were made on a PERKIN ELMER PYRIS 1 differential scanning calorimeter (D.S.C.). The calibration was done with indium and zinc. The base line was checked every day. Aluminium pans with holes were used and the samples mass was approximately 10 mg. All the temperatures measured from a peak extremum ($T_{c,n}$, T_m) are determined at less than $\pm$ 0.5°C and from a sigmoid (T_g) at less than $\pm$ 1°C. *Morphologies observations* were done with a JEOL JSM-6031 scanning electron microscope (S.E.M.).

Results and Discussion

The processing of the active CPC film is a very important step which determines many of the sensors characteristics. We have seen in a previous work [10] that the thinner the active film was the shorter the response time, so that in this study we have used two techniques known to allow obtaining thin film, spraying and spin coating. Once the deposition conditions have been optimised, pressure, speed, time, solution temperature and solutions concentration were maintained constant and only the nature of the solution, i.-e., the PS and CB was changed.

Influence of the deposition technique on electrical properties of the CPC active film: For both processes, conductivity of the samples is achieved through the percolation of CB particles resulting from the layers superposition. **Figure 3** & **Figure 4** show that the initial resistance of the sensor can be adjusted by the number of layers sprayed and also by the nature of poly(styrene) and carbon black. For sPS-CB the increase in conductivity of the sample is well correlated to an increase in specific surface of carbon black (cf. **Table 2**), but for aPS CPC, 250G gives a higher conductivity than expected from the carbon black characteristics. To explain the differences of electrical behaviour between aPS and sPS, it must be recalled that their thermo-physical characteristics are very different (cf. **Table 1**). sPS is less soluble due to its high crystallinity and melting temperature, and thus the concentration of the solution used for deposition is ten times lower than that used with aPS. More, the structuring of conductive pathways during sPS crystallisation tend to concentrate carbon black in the amorphous phases and increase conductivity compared with aPS at the same CB content. Thus it not surprising to obtain CPC films of aPS and sPS matrix with comparable resistance although shifted of about one resistance decade.

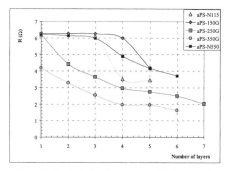

Figure 3. Resistivity decrease with the number of CPC layers for different aPS/CB associations by spraying.

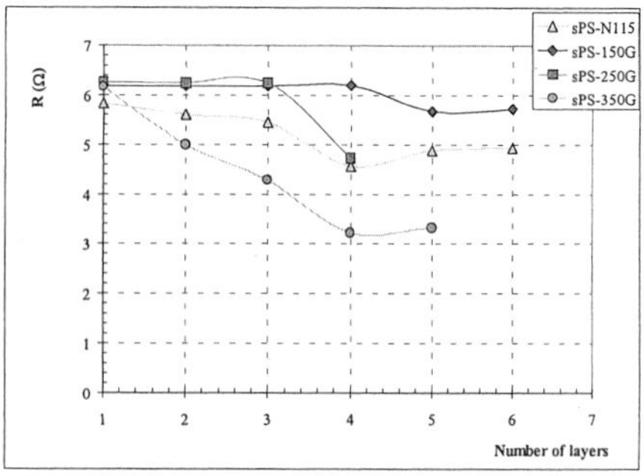

Figure 4. Resistivity decrease with the number of CPC layers for different sPS/CB associations by spraying.

Using spin coating (**Figure 5** & **Figure 6**) also allows obtaining samples with tailored resistance. The resistances of aPS CPC films are comparable to that obtained with spraying (excepted for N115 less conductive), but the resistance of sPS CPC films is much more important even with the more conductive CB (350G). More, the surface morphology shows clearly that the CB particles, denser than the polymer phase, are more sensitive to centrifugation and form large branches from the centre, clearly visible to the naked eye, which are harmful to the conducting network homogeneity. This is the reason why spraying was preferred to spin coating for the elaboration of the thin CPC active film.

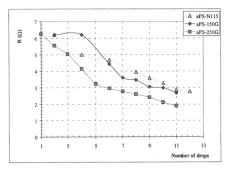

Figure 5. Resistivity decrease with the number of CPC drops for different aPS/CB associations by spin coating.

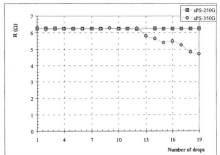

Figure 6. Resistivity decrease with the number of CPC drops for different sPS/CB associations by spin coating.

Morphology of CPC active film: The morphologies of the films obtained by spraying and for different combinations of PS and CB are presented in **Figure 7** to **Figure 14**, on the left side for aPS CPC and on the right for sPS. The comparison of **Figure 7** & **Figure 8**, **Figure 9** & **Figure 10** and **Figure 11** & **Figure 12**, respectively show that the surface morphology depends mainly on CB characteristics whereas the PS nature and the numbers of layers seem to be less important. Nevertheless, the morphology at this scale, and the related specific surface, is certainly less decisive for solvent diffusion and sensing than it is at higher magnification as illustrated by the comparison of **Figure 12** & **Figure 14**. In this later, aggregates of several thousands nanometres are clearly visible at the surface of micro pores. In fact, it is the perturbation due to solvent molecules diffusion in of such aggregates, provided that they are interconnected, that generates an electrical signal adequate for sensing.

278

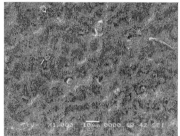

Figure 7. S.E.M. picture of aPS-N115, 2 layers.

Figure 8. S.E.M. picture of sPS-150G, 5 layers.

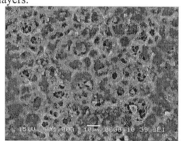

Figure 9. S.E.M. picture of aPS-250G 3 layers.

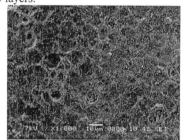

Figure 10. S.E.M. picture of sPS-250G, 3 layers.

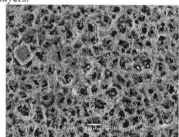

Figure 11. S.E.M. picture of aPS-350G 2 layers.

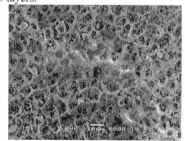

Figure 12. S.E.M. picture of sPS-350G, 3 layers.

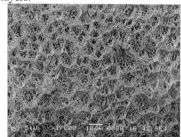

Figure 13. S.E.M. picture of aPS-N550, 5 layers.

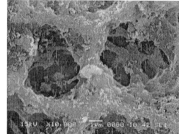

Figure 14. S.E.M. picture of sPS-350G, 3 layers.

Responses of the sensors to styrene vapours: Once the influence of the processing conditions and resulting morphologies on the sensing film electrical properties have been stated, it is finely interesting to study the impact they have on styrene vapour sensing. The response to a defined vapour can be defined by the ratio of the resistance amplitude to the initial resistance. Figure 15 shows that in the presence of 3 ppm of styrene vapour, increasing the initial resistance R_0, also increases the electrical response $\Delta R/R_0$. It is thus important to adjust R_0 during the elaboration and processing steps for each system to optimise the sensing. Another important point is to check that the electrical response is proportional to the vapour concentration. Figure 16 shows that the first results obtained for aPS-N115 in two different styrene atmospheres are promising, as even at very low vapour concentration (some ppm) the sensor is able to give a more important response at more important vapour concentration. Additional work is now in process to confirm that the responses obtained with our PS-CB sensors are quantitative for all the system studied, in a wide range of concentration and for different vapours.

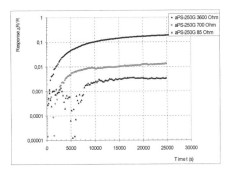

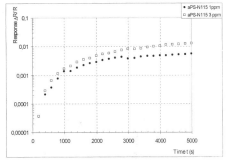

Figure 15. Responses of aPS/CB sensors to 3 ppm of styrene for different R_0 (85, 700, 3600 Ω).

Figure 16. Responses of aPS/N115 ($R_0 = 2700$ Ω) sensors to 2 styrene concentrations (1, 3 ppm).

Conclusion

Different CPC films used for vapour sensing have been elaborated from solutions by spraying and spin coating. The former process appears to be the more reproducible and adequate to design sensitive films with tuneable electrical properties. Our experiments have shown that combining aPS or sPS to five kinds of carbon black allows scanning a wide range of electrical resistance and surface morphologies to elaborate smart films for vapour sensing. For our systems an initial resistance close to 10^4 Ω appeared to give an optimal response. In these adequate conditions of elaboration and processing, our sensors gave a response for very low solvent concentration (some ppm) increasing as function of vapour concentration. Future works will concern the selectivity of these sensors to different vapours.

Acknowledgements

The authors thank M. JÉGOUSSE, F. PÉRESSE, A. BOURMAUD and N. PROBST for their contribution to this work. This project was supported by the French Ministry of Research & Technology.

[1] G. Harsanyi, "Polymer films in sensor applications: technology, materials, devices and their characteristics.", *Technomic*, Lancaster, **1995**, 113.
[2] N. Akmal, A.M. Usmani, *ACS Symposium Series*, **1998**, 690, 88.
[3] C. W. Lin, B.J. Hwang, C.R. Lee, *J. Appl. Polym. Sci.*, **1999**, 73, 2079.
[4] J. F. Feller, *J. Appl. Polym. Sci.*, **2004**, 91, 4, 2151.
[5] X. Wang, D. D. L. Chung, *Sens. & Actuators A: Phys.*, **1998**, 71, 208.
[6] E. J. Severin, B. J. Doleman, N. S. Lewis, *Anal. Chem.*, **2000**, 72, 4, 658.
[7] B. C. Munoz, G. Steinthal, S. Sunshine, *Sens. Rev.*, **1999**,19, 4, 300.
[8] B. Lundberg, B. Sundquist, *J. Appl. Phys.*, **1986**, 60, 1074.
[9] J.F. Feller, Y. Grohens, *Sens. & Actuators B: Chem.*, **2004**, 97, 2-3, 231.
[10] J. F. Feller, D. Langevin, S. Marais, *Synth. Met.*, **2004**, 144, 1, 81.

Macromol. Symp. **2005**, *222*, 281-286

Ordering in Bio-Polyelectrolyte Chitosan Solutions

Laurent David,[1] *Alexandra Montembault,*[1] *Nadège Vizio,*[1] *Agnès Crépet,*[1] *Christophe Viton,*[1] *Alain Domard,*[1] *Isabelle Morfin,*[2] *Cyrille Rochas*[2]

[1]Laboratoire des Matériaux Polymères et Biomatériaux, Université Claude Bernard, UMR CNRS 5627 "Ingéniérie des Matériaux Polymères", 15 bd. Latarjet, Bât. ISTIL, 69622 Villeurbanne Cedex, France
E-mail: laurent.david@univ-lyon1.fr
[2]Laboratoire de Spectrométrie Physique, Université Joseph Fourier, UMR CNRS 5588, 140 avenue de la Physique, B.P. 87, 38402 Saint Martin d'Hères, Cedex, France

Summary: The scaling of the polyelectrolyte scattering peak in chitosan solutions, as deduced from the relation $q_{max} \sim c_p^{\alpha}$ was studied by synchrotron SAXS as a function of the charge density of the polymer. We observe a variation in the α exponent corresponding to the limit of the ionic condensation, by varying the degree of acetylation of the polymer. The nature of the solution medium also affects the polyelectrolyte peak, and it is shown that in alcoholic/water mixtures, the lower dissociation of the acid induces a lower charge density, thus influencing the polyelectrolyte ordering.

Keywords: chitosan; lattice model; polyelectrolyte; SAXS

Introduction.

Chitosan is obtained by chemical modification (deacetylation) of chitin, one of the most widespread natural polymers with cellulose. Both polymers exhibit the same chemical structure, as they are linear copolymers of N-acetyl-D-glucosamine and D-glusoamine with β,(1→4) glycosidic linkages. Chitosan is obtained when the Degree of Acetylation (DA^1) is below 60%. This results into a soluble polyelectrolyte polymer in diluted acidic solutions at pH<6, well below the pK_o of the NH_3^+ moieties. Chitosan can also be obtained in other physical forms such as gels [1,2], soft and hard solids [2], bulk samples, films, fibres[3] and also more complex structures with macroscopic property gradients [4]. At last, this polymer is now well-known for its biological interest with interesting antumoral[5], antifungal[6] and antibacterial[7] activity, but mostly for bioactivity with applications for wound healing in various soft and hard tissues[8,9].

Our aim is to gain insight into the structure and microstructure of the chitosan polymers in order to better explain their macroscopic properties in their various physical states,

---

[1] In the experimental conditions of our study (pH<6), the fraction of charged monomer *f* in chitosan polyelectrolyte polymer is related to the Degree of Acetylation *DA* by *f*=1-*DA*.

 DOI: 10.1002/masy.200550437

together with their biological features. This work is devoted to chitosan in water and alcoholic solutions in salt-free conditions.

Experimental

<u>Materials</u>

Two weakly acetylated chitosans were used. They were produced from squid pens and purchased from France Chitine (batch 11.03, DA ~2.6%, M_w ~540000 g.mol^{-1} and batch 114, DA ~5.2%, M_w ~530000 g.mol^{-1}). Prior to use, the polymers were purified according to[10]. All other samples (DA=36.7; DA=46.2 and DA=65.5%) were prepared by homogeneous re-acetylation of the chitosan of an initial DA=5.2 % with acetic anhydride in a water/alcohol solution[11].

<u>Experimental techniques</u>

We studied the so-called 'polyelectrolyte peak' by small angle synchrotron X-ray scattering (ESRF-Grenoble France BM2-D2AM beamline). A synchrotron source was required for our study because the intensity scattered by the polyelectrolyte solution is not measurable by conventional sources.

The data were collected at an incident photon energy of 16 keV. We used a bi-dimensional detector (CCD camera from *Ropper Scientific*). All the data corrections were performed by the software *bm2img* available on D2AM beamline. The data were corrected from the dark and flat field responses of the detector. Moreover, the background (cell filled with water) was subtracted. The distortion introduced by the optical fiber bundle was also corrected. Lastly, the radial average around the image center (location of the center of the incident beam) was performed for the q-range calibration standard (silver behenate) and the chitosan solutions.

The viscous solutions were introduced in Low Density PolyEthylene (LDPE) cylindrical sample holders (internal diameter~5mm) with two holes perpendicular to the main axis, closed by adhesive Kapton in order to avoid the scattering contribution of LDPE.

Results

In all studied solutions, the scattering intensity could be separated into (*i*) a low angle contribution arising from long range electronic density fluctuations and associated with chain aggregation and (*ii*) the 'polyelectrolyte peak' itself. We studied the shape of the scattering curve from the phenomenological relation accounting for both scattering components:

$$I(q) = \frac{C}{q^\gamma} + \frac{B}{4\left[\dfrac{q - q_{max}}{w}\right]^2} \tag{1}$$

This equation is very similar to that proposed by Wang[12], and only differs by an additionnal constant term that will be not used here. In equation 1, q_{max} is the location of the Lorentzian peak, w is the full width at half maximum, and B is the intensity of the polyelectrolyte peak. We thus consider that the value of q_{max} is close to $2\pi/d$ where d is the most probable interchain distance. An exemple of modelling is given in figure 1. Although we observed some evolutions of the parameter C and γ describing the contribution of the long range fluctuations, we will only report on the parameters associated with the polyelectrolyte peak. The dependence of q_{max} with the polymer concentration c_p will be described, as usual, by the scaling relation:

$$q_{max} = q_0 . c_p^{\;\alpha} \tag{2}$$

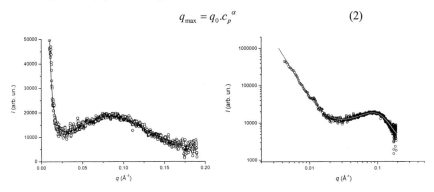

Figure 1. Example of modeling of a scattering curve according to equation 1 (solid line with C=0.045 arb.un. Å$^\gamma$; γ=3; B=18500 arb.un. Å^2; q_{max}=0.085 Å^{-1}; w=0.13 Å^{-1}) and experimental data (o) for chitosan acetate aqueous solution, with DA=5.2% and polymer concentration c_p=0.015 g/g. Left: linear representation, Right: log-log representation.

Influence of DA on the polyelectrolyte peak

The concentration dependence of the location of the polyelectrolyte peak, for different DA values was characterized in the case of chitosan acetate in water. The power law (equation 2) is well adapted to describe the results in a large concentration range. The power laws observed for low DA (*i.e.* at 2.6 and 5.2%) exhibit a higher exponent than for the high DA

values (46.2% and 65.5%). The values of the scaling exponent α are given by figure 2 as a function of the degree of acetylation.

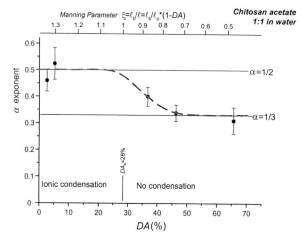

Figure 2. Change in the α exponent of the scaling relation (2), as a function of DA for chitosan acetate in water. The transition DA between $\alpha \sim 1/2$ to $\alpha \sim 1/3$ is close to the ionic condensation limit, as predicted by Manning theory.

The transition observed from low to high DA bears many similarities with that reported by Baigl[13] in the case of partly sulfonated polystyrene. The transition looks continuous, and the polyelectrolyte peak is broader close to the ionic condensation[14] limit, at $DA \sim 28\%$[15] as shown on figure 3. Indeed the Manning parameter is theoretically related to the DA value by $\xi = [\ell_B/\ell_o](1-DA)$ where ℓ_B is the Bjerrum length (7.1 Å in water) and ℓ_o is the length of the repeating unit (5.1 Å in the case of chitosan).

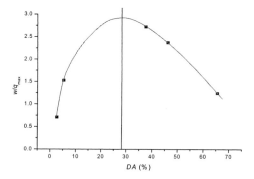

Figure 3. Evolution of the breadth parameter of the polyelectrolyte peak, w/q_{max} as a function of DA for chitosan acetate in water at $c_p = 1.5\%$

We can interpret these data as the transition between a structure consisting in a collection of parallel polyelectrolyte rod-like chains (with $d\sim c_p^{-1/2}$)[16] at low DA when ionic condensation is present, to a solution with punctual electrostatic interaction sites (with $d\sim c_p^{-1/3}$) when the apparent charge density of the polymer decreases, escaping from ionic condensation. In the DA range associated with a Manning parameter ξ close to unity, the coexistance of both structural regimes can explain the increase of the breadth of the polyelectrolyte peak.

<u>Influence of the solution medium on the polyelectrolyte peak</u>

For a very low DA chitosan (2.6%), addition of 1,2 propanediol to the solution decreased the amplitude of the polyelectrolyte peak and shifted the polyelectrolyte peak to larger distances (see figure 4). Moreover, the breadth parameter of the polyelectrolyte peak was largely increased in the high q range, so that the Lorentzian modeling is not adequate in this case. At higher alcohol concentrations, we observed a disappearance of the polyelectrolyte peak and the formation of a gel. An evolution of the dielectric constant of the solution medium alone can not be invoked to explain the results: the decrease of the relative dielectric constant form 80 in water to 27.5 in pure alcoholic medium leads to a larger Bjerrum length $\ell_B = \dfrac{q^2}{\varepsilon kT}$ and thus should promote Manning condensation ordering in the polyelectrolyte solution. As a result, we interpret our results in terms of a decrease of apparent density of charge of the chains, since the dissociation of acetic acid is lower in hydro-alcoholic media.

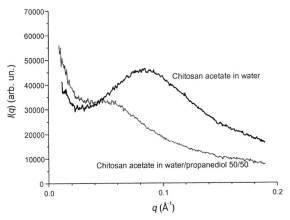

Figure 4. Evolution of the polyelectrolyte peaks for chitosan acetate in water (black line) and in water/propanediol for a concentration of 1.5% and a DA of 2.6%

Conclusion

Chitosan is a versatile and unique natural cationic polyelectrolyte, enabling the fundamental study of polyelectrolyte solutions ordering in various experimental conditions. The role of *DA* is to change the apparent charge density that in turn controls the structure of chitosan solutions. We observed the polyelectrolyte peak in the ionic condensation regime and out of the condensation regime: the scaling exponent α yielding the location of the polyelectrolyte peak q_{max} as a function of polymer concentration changes from 1/2 to 1/3 by increasing the degree of acetylation from below 28% to above 28%, *i.e.* the Manning parameter from above 1 to below 1. This transition[13] is rather gradual, and in the vicinity of the limit between the two ordering regimes, the width of the polyelectrolyte peak is the largest, as a result of the coexistence of different arrangement types. Such an effect is qualitatively observed if the solution medium is enriched in propanediol, thus inducing a lower dissociation of the acid, and thus, again, a lower charged fraction of the polyelectrolyte.

[1] Thesis, A. Montembault, UCB Lyon1 2004.
[2] Thesis, N. Vizio, UCB Lyon1 2005.
[3] Thesis, L. Notin, UCB Lyon1 2005.
[4] Thesis, S. Ladet, UCB Lyon1 2006.
[5] H.O. Pae, W.G. Seo, N.Y. Kim, G.S. Oh, G.E. Kim, Y.H. Kim, H.J. Kwak, Y.G. Yun, C.D. Jun, H.T. Chunhg, *Leukemia Research*, **2001**, *25*, 339-346.
[6] C. Jarry, C. Chaput, A. Chenite, M.A. Renaud, M. Bushmann, J.C. Leroux, *J. Biom. Mat. Res.*, **2001**, *58*, 127-135.
[7] Y. Shin, D.I. Yoo, K.Min, *J. Appl. Polym. Sci.*, **1999**, *74*, 2911-2916.
[8] J.K. Francis Suh, H.W.T. Mattew, *Biomaterials*, **2000**, *21*, 2589-2598.
[9] C. Muzzarelli, R.A. Muzzarelli, *J. Inorg. Biochem.* **2002**, 92, 89-94.
[10] C. Schatz, C. Viton, T. Delair, C. Pichot, A Domard,. *Biomacromolecules*, **2003**, *4*, 641.
[11] L.Vachoud, N.Zydowicz , A. Domard, *Carbohydr. Res.* **1997**, 302, 169.
[12] D. Wang, J. Lal, D. Moses, G.C. Bazan, AJ. Heeger, *Chem. Phys. Lett.*, **2001**, *348*, 411-415.
[13] D. Baigl, R. Ober, D. Qu, A. Fery, C.E. Williams *Europhys. Letters*, **2003** , *62(4)*, 588-594.
D. Qu, D. Baigl, C. E. Williams, H. Möhwald, A. Fery, *Macromolecules*, **2003**, *36*, 6878-6883.
W. Essafi, F. Lafuma, C.E. Williams, *J. Phys. France* **1995**, *5*, 1269-1275.
[14] G. S. Manning, *Biophysical Chemistry*, **1977**, *7*, 95-102.
G. S. Manning, *Physica A* **1996**, *231*, 236-253.
[15] C. Schatz, J.M. Lucas, A. Domard, C. Pichot, C. Viton, T. Delair, *Langmuir*, **2004**, *18*, in press
[16] S. Lifson, A. Katchalsky, *J. Polym. Sci.*, **1954**, *13*, 43.

Construction of a Low Cost Photoacoustic Spectrometer for Characterization of Materials

M. A. Jothi Rajan,[*1,2] *Arockiam Thaddeus,*[3] *T. Mathavan,*[4]
T. S. Vivekanandam,[5] *S. Umapathy*[1]

[1] School of Physics, Madurai Kamaraj University, Madurai –625021, India
Fax: (+91) 4549 287208; E-mail: anjellojothi@rediffmail.com
[2] Department of Physics, Arul Anandar College, Karumathur – 625514, India
[3] P.G Dept. of Zoology, JayarajAnnapackiam College, Periyakulam –625601, India
[4] P.G Dept. of Physics, NMSSVN College, Nagamalai –625019, India
[5] School of Chemistry, Madurai Kamaraj University, Madurai –625021, India

Summary: During the past few years, another optical technique has been developed to study those materials, which cannot be studied, by the conventional transmission or reflection techniques. The present technique called *Photoacoustic spectroscopy* or PAS is different from the conventional techniques chiefly in that the interaction of the incident energy of the photons with the materials under investigation is studied not through subsequent detection and analysis of some of the photons, but rather through a direct measure of the energy absorbed by the material. The aim of this presentation is to highlight the construction of a simple *Photoacoustic spectrometer* which can easily be constructed even in high school and college laboratories with the available low cost but efficient components and use it for characterization of solid (opaque or transparent), liquid and gas samples under investigations. The essential parts of the photoacoustic spectrometer designed in the laboratory (MADURAI - PA SPECTROMETER), consists of three main components.The total cost comes around 900 Euros. It is an affordable cost for researchers working with paucity of funds and facilities and many constraints especially in the developing countries. In the next few years we aim to study material characterization using MADURAI –PA SPECTROMETER.

Keywords: low cost spectrometer; photoacoustics; photoacoustic spectrometer

Introduction

Optical spectroscopy has been a scientific tool for over a century and it has proven invaluable in studies on reasonably clear media such as solutions and crystals and on specularly reflective surfaces. There are, however several instances where conventional transmission spectroscopy is inadequate even for the case of clear transparent materials. Such situations arise when one is attempting to measure very weak absorption. In

DOI: 10.1002/masy.200550438

addition to weakly absorbing materials there are a great many non-gaseous substances, both organic and inorganic, that are not readily amenable to the conventional transmission or reflection modes of optical spectroscopy. These are usually light scattering materials; such as powders, amorphous solids, gels, smears, suspensions and nanoparticles. To fill this gap, photoacoustic spectroscopy has come into existence. Photoacoustics (PA) is well known in research but it is less known to Physics students at undergraduate and postgraduate levels. In this article a simple and low-cost experimental design of the photo acoustic spectrometer is explained which can be easily designed in any developing laboratories.

The principle of photoacoustic is the generation of acoustic energy from modulated light energy. The resulting energy propagates away from the source as acoustic waves. That is, the photoacoustic effect is the generation of acoustic waves in a sample resulting from the absorption of photons. The sample may be solid, liquid, gas, powder, gel, thin films or nanoparticles. Alexander Graham Bell [1] discovered the photoacoustic phenomena in the year 1880. The PA effect is represented pictorially in Figure 1.

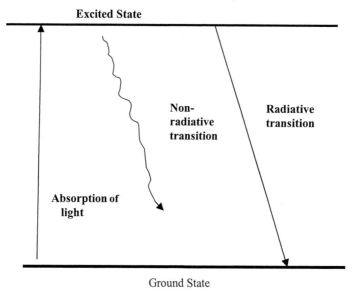

Figure 1. Origin of the photoacoustic spectrum.

Experimental

Construction of PA spectrometer

All the parts of the PA spectrometer were designed and assembled in School of Physics of Madurai Kamaraj University. The only readymade instrument was the digital storage oscilloscope, which was used to make measurements. The block diagram of the so-called 'MADURAI-PA SPECTROMETER' is shown in Figure 2.

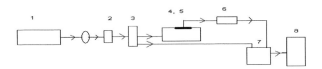

Figure 2. Block diagram of Madurai PA spectrometer; 1.Source; 2.Monochromator; 3. Chopper; 4. PA cell; 5. Microphone; 6. Pre-amplifier; 7. Lock-in amplifier; 8. Digital Storage Oscilloscope.

Source

The source is a 1000 watt tungsten halogen lamp [2] with the reflecting mirror and condensing lens arrangement, which can be moved, collimated the beam of light back and forth for convenience of focusing. A sturdy transformer was assembled to operate tungsten halogen lamp at 12 amperes a.c current and was tested for its performance.

Monochromator

Since the cost of monochromator is high we employed a simple procedure. Since it was intended to carry out the experiments in the visible region of the electromagnetic spectrum, we used colored filter papers (highly transparent) violet, indigo, blue, green, yellow, orange and red. The appropriate filter paper was pasted to a lens holder, which will act as a monochromator when white light passes through it. Care was taken to place the colored filter paper intact and clean.

Chopper

We have made the beam chopper from thick aluminum sheets. The aluminum sheets were cut accordingly so that it will convenient to vary the frequency of the chopper. Hence we cut out three blades of suitable dimensions. The shape of the chopper was that of a fan blade except that the blades are plane instead of twisted. Like this we have designed the different shopper with four blades and two blades. The number of blades increases the frequency of the chopping. This chopping blade with proper arrangements for rotations is connected to an a.c motor. The operating voltage for the a.c motor can be varied from 0-230V. Change in the motor speed changes the chopping frequency. This can be adjusted by changing the power supply voltage, which is variable from 0V to 230V. According to the required experimental frequency a proper chopper blade was selected from the 3 chopping blades. The chopper was placed near or at the focus of the lens. The shape of one of the chopping blades is shown in Figure.3.

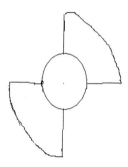

Figure 3. Two segments chopper blade.

Photoacoustic cell

The photoacoustic cell is one of the most important components that require skillful designing suitable for obtaining PA spectra according to the nature of the sample (solid, liquid, gas). For solid samples we have designed a PA cell as shown in figure 4. The PA cell is made up of a flint glass funnel whose larger diameter is 10 cm. A funnel of high thermal conductivity material such as metal may result in a weaker signal. Therefore glass funnel was used in our experiment. A cylindrical glass vessel to which liquid can be circulated and the sample in the cell can be kept at any desired temperature by connecting inlet and outlet to the constant temperature bath envelops the whole funnel. The interior of the glass funnel is coated with the lampblack. When opaque layer of carbon has been deposited over the entire interior of the funnel it acts as a backing material. The large opening of a funnel is sealed with watch glass using best adhesive paste. In the curved watch glass the convex side should be placed inside the funnel to reduce volume of the shield chamber and subsequently focusing of the beam. A thin sample holder made of brass/aluminum is placed inside the stem of the glass funnel and in front of the holder a sensitive microphone is placed perpendicular to a sample holder but without touching it.

Thin electrical connecting wires are taken from the microphone without disturbing the airtight arrangement of the experimental PA cell setup. One of the detectors namely the microphone is placed inside the funnel which was inevitable. All the other detector devices are assembled in a separate chassis.

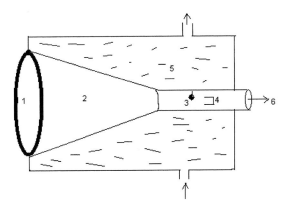

Figure 4. PA cell for solid samples; 1. curved watch glass; 2. lampblack coating; 3. microphone; 4. sample holder; 5. thermostat; 6. PA signal to detectors.

For liquid samples we have designed a unique PA cell as shown in figure 5. The PA cell is made up of a small glass funnel. In the center of the stem of the funnel and air tight compartment is drilled to insert a quartz container to hold the liquid sample. The microphone can be placed over the sample without touching it. The interior of the cell is coated with lampblack except the region containing the quartz cell. The entire is made airtight. The large opening of the funnel is sealed with the clean watch glass. For gaseous samples we designed a innovative PA cell as shown in Figure 6. The cell is cylindrical in shape and there is provision for heating the gaseous sample at any desired temperature by externally circulating liquid system. For PA students of gaseous samples the entire cell should fill the gas and no sample holder is required. The interior of the cell is uniformly coated with lampblack. One of the ends of the cylindrical cell is completely sealed and is blackened inside with lampblack and the microphone is sealed inside the glass vessel.

The other end of the cylindrical glass vessel has an opening and a sliding door made of quartz glass will be sealing the entire cell.

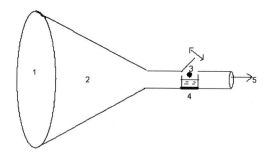

Figure 5. PA cell for liquid samples; 1.curved watch glass; 2. lampblack coating; 3. microphone; 4. quartz sample holder; 5. PA signal to detectors.

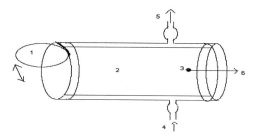

Figure 6. PA cell for gaseous samples; 1.quartz glass window; 2. lampblack coating; 3. microphone; 4. thermostat inlet; 5. thermostat outlet; 6. PA signal to detectors.

Detector

In all the three cells the two terminals of the microphone are projecting outside so that connections can be made easily without disturbing the PA cell. The microphone is connected to an external bias of 2V d.c supply. The other end of the microphone is used as input for the amplifier, which is designed with an IC 741 chip. A very simple and low cost IC 741 operational amplifier with negative feedback is used and the gain is maintained to be 100. This amplifier circuit is shown in figure 7. The power supplies that are needed for this amplifier are constructed with IC 7812 and IC 7912. The lock-in amplifier set up is made p of indigenous electronic components so that it selects only the audio signal, which is at the chopping frequency. A reference signal from the chopper is given to the lock-in amplifier. In the same lock-in amplifier set up which is kept in an iron chassis three digital displays are incorporated to measure chopping frequency, photoacoustic signal amplitude and phase of the photoacoustic signal. Provisions are made to give the output to the digital storage oscilloscope to trace the waveform and to check the PA signal amplitude and phase read by the digital displays are one and the same. In our experiment we used the digital panel meter to find the chopping frequency. The digital storage oscilloscope (DSO, 20MHz, GOULD) measured the PA signal amplitude and phase.

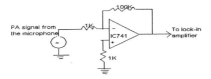

Figure 7. Circuit diagram of pre-amplifier.

Results and Discussion

To check the performance of the constructed "MADURAI - PA SPECTROMETER", we conducted experiments with a known solid sample, liquid sample and gaseous sample using the three different cells. The results that were obtained are encouraging and fall within an error of ±3% from the other references.

Solid sample

By keeping the wavelength fixed (green color filter was used) the depth profile analysis was carried out and the corresponding PA signal is measured by varying the chopping frequency. The sample that was tested was solid poly(methyl acrylate) synthesized in our laboratory. The characteristic frequency f_c was found out. This frequency is that point at which the sample goes from thermally thin to thermally thick state [3]. The thermal diffusivity of the sample was calculated and found to be $1.9 \times 10^{-6} m^2 sec^{-1}$. Jothi Rajan et al. [4] have reported for the same PMA sample with a sophisticated and very costly photoacoustic spectrometer (EG&G MODELS) the diffusivity value of $1.8 \times 10^{-6} m^2 sec^{-1}$. Thus the indigenously designed low cost set up is in no way inferior to the one that was imported.

Liquid sample

To study the thermal diffusivity of polyaniline in N-methylpyrrolidone (NMP) we used the liquid cell and followed the depth profile analysis and found the characteristic frequency. Hence we found thermal diffusivity to be $25.6 \times 10^{-6} m^2 sec^{-1}$. Pilla et al. [5] have reported the thermal diffusivity value for polyvinylacetate /polyaniline solution by thermal measurements as $1.05 \times 10^{-7} m^2 sec^{-1}$. This measurement is also fairly good and acceptable within the error limits.

Gaseous sample

We used the gas PA cell to study the photochemical deexcitation of NO_2 gas at a pressure of 10 torr. Here we followed the wavelength scanning method. The chopping frequency is kept fixed and the wavelength is changed and accordingly the PA signal amplitude is measured. Harshbarger and Robin [6] have got a similar type of spectra at a chopping frequency of 510 Hz. In our experiment the chopping frequency was fixed at 525 Hz. The spectra are compared are shown in figure 8. These are in good agreement with our results within the acceptable error limits.

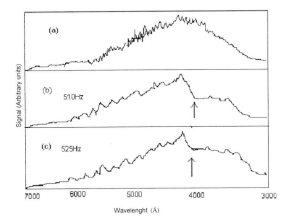

Figure 8. Comparison of the (a) optical and (b) photoacoustic spectra of NO$_2$ at 10 torr (c) photoacoustic spectra of NO$_2$ at 10 torr in the present work.

Conclusions

The indigenously and economically designed MADURAI - PA SPECTROMETER will be used for our future studies on nanocomposites of organic and inorganic polymer and biological samples. Since the PA technique is a nondestructive testing and evaluating tool with maximum facilities for obtaining best results of rare samples we intend using it and the same also fits within the limits of our funding. This technique gives fairly accurate results as any other conventional methods. This simple spectrometer can be constructed in college laboratories and number of innovative experiments can be carried out. The total cost of the entire set up came around 900 Euros.

In future we aim at using the photopyroelectric nature of the same phenomena. The designing of the low cost photopyroelectric spectrometer is in progress in our laboratory with the limited funding. We intend testing biological samples also by this technique and explore the various causes for the spread of viruses between human and animals.

Acknowledgements

One of the authors M.A.Jothi Rajan acknowledges the University Grants Commission of India for awarding FIP fellowship to carryout this work.

[1]. A.G. Bell, *Philos. Mag.*, **1881**, <u>11</u>, 510.

[2]. P. Ganguly, C.N.R. Rao, *Proc. Indian Acad. Sci. (Chem. Sci.)*, June**1981**, <u>90(3)</u>, 153-214.

[3]. A. Rosencwaig, *J. Appl. Phys.*, **1978**, <u>A49</u>, 2905.

[4]. M.A. Jothi Rajan, T.S. Vivekanandam, S.K. Ramakrishnan, K. Ramachandran, S. Umapathy, *J. Appl. Polym. Sci.*, **2004**, <u>93(3)</u>,1071-1076.

[5] V. Pilla, DT. Balogh, RM. Faria, T. Catunda, *Review of Sci. Ins.*, January **2003**, <u>4(1)</u>, part 2, 866-868.

[6]. W.R.Harshbarger, M.B. Robin, *Acc. Chem. Res.*, **1973a**, <u>6</u>, 329.